Collins

INTERNATIONAL
LOWER SECONDARY

Computing
Teacher's Guide

7

Rebecca Franks, Dr Tracy Gardner and Liz Smart

William Collins' dream of knowledge for all began with the publication of his first book in 1819.

A self-educated mill worker, he not only enriched millions of lives, but also founded a flourishing publishing house. Today, staying true to this spirit, Collins books are packed with inspiration, innovation and practical expertise. They place you at the centre of a world of possibility and give you exactly what you need to explore it.

Collins. Freedom to teach

Published by Collins

An imprint of HarperCollins*Publishers*
The News Building, 1 London Bridge Street, London, SE1 9GF, UK

HarperCollinsPublishers
Macken House, 39/40 Mayor Street Upper, Dublin 1, D01 C9W8, Ireland

Browse the complete Collins catalogue at collins.co.uk

10 9 8 7 6 5 4 3 2 1

ISBN 978-0-00-868408-2

British Library Cataloguing-in-Publication Data
A catalogue record for this publication is available from the British Library.

Authors: Rebecca Franks, Dr Tracy Gardner and Liz Smart
Publisher: Catherine Martin
Product manager: Saaleh Patel
Project manager: Just Content Ltd
Development editor: Julie Bond
Copyeditor: Becca Law
Proofreader: Laura Connell and Tim Jackson
Cover designer: Gordon McGilp
Cover illustrator: Amparo Barrera, Kneath Associates
Internal designer: Ken Vail Graphic Design
Typesetter: Sam Vail
Production controller: Lyndsey Rogers
Printed and bound by Ashford Colour Press

FSC www.fsc.org MIX Paper | Supporting responsible forestry FSC™ C007454

This book contains FSC™ certified paper and other controlled sources to ensure responsible forest management.

For more information visit: www.harpercollins.co.uk/green

Acknowledgements

Support materials and screenshots are licensed under the Creative Commons Attribution-ShareAlike 2.0 license. We are grateful to the following for permission to reproduce screenshots. In some instances, we have been unable to trace the owners of copyright material, and we would appreciate any information that would enable us to do so.

Python Software Foundation: Permission obtained to use screenshots demonstrating Python programming language features.

Scratch Foundation: Authorised usage of screenshots showcasing Scratch programming environment elements.

EduBlocks: Granted clearance for the inclusion of screenshots depicting EduBlocks interface and functionalities.

Scratch is developed by the Lifelong Kindergarten Group at the MIT Media Lab.

PSF's License Agreement and PSF's notice of copyright, i.e., "Copyright © 2001 Python Software Foundation; All Rights Reserved" are retained in Python 2.0.1 alone or in any derivative version prepared by Collins (Licensee).

Contents

Teacher's Guide

Introduction

Stage 7 Computing extends the robust foundation laid in Stages 1 to 6 of the Collins course. Students delve into the safe and respectful utilisation of online collaboration tools. Engaging practical activities guide them through the exploration of new machine learning and artificial intelligence concepts. Enhancing their software development skills, effective use of flowcharts and prototyping is emphasised. Binary representation of data is studied, leading to the collaborative creation of an art installation. Technical aspects of networking, including encryption, are also explored.

Overview

The Collins series is built around six themes relating to computing and digital literacy. These are:

- Our digital world – Providing the tools and insights into safely navigating the digital world around us
- Content creation – Creating content using a variety of software, from office tools to video production
- Create with code – Providing the fundamentals of programming and computational thinking skills
- How computers work – Lifting the lid on the specific technologies that make up a computer system
- Connect the world – Understanding how the world is connected through networks, the internet and the World Wide Web
- The power of data – Collecting, analysing and presenting data linked to real-world activities that create or empower change

Stage 7 has eight chapters that relate directly to these six themes, with three chapters dedicated to creating with code.

Stage 7 Kit list

Chapter 1 – Our digital world	• This unit covers illegal online activity, such as cyberbullying and child exploitation. *Make sure that you are familiar with the laws, procedures and safeguarding guidance in your local area prior to teaching this unit.* • Computers with access to the Scratch website • Microphones • Computers with access to the Machine Learning for Kids website
Chapter 2 – Content creation	• An online word processor with collaborative tools enabled • A video of a local news report that is in interview style • Video recording equipment • Web-enabled video recording equipment • 3 × video recordings of different lengths to demonstrate how file size is impacted by length of recording • Projector screen
Chapter 3 – Create	• Playing cards

with code 1	• Computers with access to the Scratch website • The ability to save and load projects in Scratch
Chapter 4 – How computers work	• Access to the Scratch website available on your teacher machine • A collection of images that relate to your local community • Sticky tape or a display board with pins • Coloured pencils (optional) • Squared paper (optional)
Chapter 5 – Create with code 2	• Computers with access to the EduBlocks platform • Printed copies (or digital access) to the EduBlocks 123 Coding Cards which map Scratch to EduBlocks to Python. (Website IntR1)
Chapter 6 – Connect the world	• Access to computers that have access to the Scratch website • Locations of local cell towers or masts (optional) • Access to a colour printer (optional) (alternative: a way to share and annotate digital images)
Chapter 7 – The power of data	• Computers with access to a search engine, the Scratch website, form creation tools and spreadsheet software
Chapter 8 – Create with code 3	• Computers that have access to the EduBlocks classroom • Enough micro:bits for at least paired work but ideally for one each • Coloured pencils (optional)

Pedagogy

Our overall approach

Our approach to teaching is based on the latest research in ensuring that computing classrooms are inclusive environments that foster greater participation of underrepresented groups in computing and technology.

Each chapter is carefully organised to develop essential knowledge and skills whilst working towards the creation of a final project. The final projects are designed to boost creative skills and give students the opportunity to make decisions and develop outputs that matter to them. All chapters end with a showcase lesson that gives students the opportunity to develop their presenting skills, whilst gaining valuable feedback on their work.

Every lesson plan states the key pedagogy that is used in that lesson. Stage 7 uses the following key pedagogies:

Classroom talk
Classroom talk is the communication that happens in a classroom, between teacher and student, student and teacher, and between students. Discussion and social interaction plays an important role in computing education. Throughout every lesson in Stage 7 you will find 'Discuss' activities. You are encouraged to incorporate topics that have been generated by students and encourage them to share their ideas and perspectives and to ask any questions they have.

Benefits:
- Classroom talk provides an opportunity for multi-way knowledge sharing and building a shared understanding.
- Valuing student voices and perspectives increases their sense of connection to computing.

Jenkins, 2017. Classroom Talk and Computational Thinking DOI: (Website IntR2)

Get creative
This key pedagogy is used to encourage students to be creative with code or technology and provides them with the freedom to make their own design choices. Stage 7 provides a wealth of opportunities to get creative with technology, from contributing to a piece of collaborative pixel art to building a 'classroom welcome' device.

Benefits:
- Students create projects that are culturally relevant to them, leading to greater engagement.
- Providing more choice in the projects that students create leads to a greater sense of agency, particularly amongst girls.
- Spending more time on the creative elements of computing leads to greater engagement of underrepresented groups.

Steve Human et al, July 2022. (Website IntR3).
Sue Sentance et al, August 2021. (Website IntR4)

Real-world contexts
Lessons make regular links to everyday life scenarios where students may encounter the concepts they are learning about. Links are also made to other curriculum topics. In Stage 7, students learn about how AI is used in their everyday lives and how simulators are used in industry.

Benefits:
- Using real-world contexts aids students' understanding by connecting computing concepts to their everyday lives.
- Real-world contexts demonstrate the value of computing, which engages a more diverse group of students.
- Real-world contexts show the practical application of computing and make links to the world of work.

Unplugged
Unplugged activities demonstrate computing concepts without using computers. In Stage 7, students create a pixel art image in an unplugged activity.

Benefits:
- Unplugged activities have been demonstrated to positively affect students computational thinking skills.

- Unplugged activities can increase students' engagement with concepts and develop deeper understanding.
- Unplugged activities can be helpful where the amount of time that students have access to computers is limited.

Chen et al, 2023. Fostering computational thinking through unplugged activities: A systematic literature review and meta-analysis (Website IntR5).

Develop practical skills

To be effective in computing, students need to develop practical skills such as using a keyboard and mouse and using IT tools.

Benefits:

- Developing practical skills enables students to become effective users of technology.
- Developing practical skills means that students have more capacity to focus on the new computing concepts they are learning.

Showcase

Showcasing is presenting a project to an audience. In Stage 7, every chapter ends with a showcase of the work that the students have completed.

Benefits:

- Showcasing requires students to reflect on their work so that they can explain it to others.
- Showcasing develops important communication skills.
- Showcasing enables students in the audience to develop the ability to give feedback.
- Showcasing provides an opportunity to celebrate the achievements of students and the tangible outcomes of computing work.

Building on prior knowledge

Building on prior knowledge involves recall of existing knowledge of programming when learning a new programming language. Similarities can be beneficial to conceptual transfer, though attention must be paid to explaining differences between programming languages. In Stage 7, students use their knowledge of Scratch and EduBlocks as they learn the Python text-based programming language.

Benefits:

- It enables students to acquire new knowledge of programming concepts more rapidly by building on existing knowledge (attention must be paid to differences to avoid confusion).
- It makes new programming languages less intimidating due to familiarity with similar concepts.

Ethel Tshukudu et al, 2020. Understanding Conceptual Transfer for Students Learning New Programming Languages. (Website IntR6).

Investigate

Investigate is the 'I' in the PRIMM (Predict, Run, Investigate, Modify and Make) methodology for learning programming. It involves students interacting with code in a variety of ways to deepen their understanding. In Stage 7, students carry out exploratory activities and answer questions to develop their understanding of new programming concepts.

Benefits:

- Investigation builds deeper understanding of coding concepts through interacting with code in different ways.
- It promotes discussion of questions that arise when interacting with code.

Sue Sentance et al, 2019. Teachers' Experiences of using PRIMM to Teach Programming in School. (Website IntR7).

Assessment

The books are designed to follow a mastery learning approach where all learners are supported and challenged to meet the age related expectations of the curriculum. Formative assessment should be used regularly to check the overall understanding of your class. Each chapter has a final project that is assessed summatively. Appropriate scaffolding is provided to allow for the majority of learners to meet expectations.

Formative assessment in the student workbook

The student workbook provides opportunities for learners to apply their learning to various activities. The outcome of these tasks can be used as part of a formative assessment process to measure the learner progress throughout the chapter. Where a workbook question has a specific answer, answers have been provided. There are also 'model answers' to guide your assessment needs.

Formative assessment and the lesson plan guidance

Each lesson plan has a guidance column. Where appropriate, additional support has been provided for the type of answers that you are looking for when questioning the learning. This will help to guide your formative assessment. If learners are struggling to understand the concept then you should revisit it with the support of the scaffolding and guidance provided.

Summative assessment and the project rubrics

Every chapter has a final project that the learners are working towards. Each project has a project brief with requirements for learners to meet. The projects are then summatively assessed using a rubric. Each rubric has three columns 'Working towards', 'Meets expected', and 'Exceeding expected'. The majority of the class should be able to achieve the 'Meets expected' criteria. Some learners will exceed this expectation and go a step further than required by the project brief. A few learners may fall into the 'Working towards' category and should receive additional support and scaffolding.

Learners should have access to the rubric at the start of a project so that it is clear how they will be assessed. This also helps to support their metacognitive skills.

Summative assessment and showcasing

Each chapter ends with a showcase which provides an opportunity for summative assessment. Teachers can complete the assessment rubric during showcase lessons. Teachers can ask additional questions of learners to establish their understanding of specific criteria.

Differentiation

Projects are differentiated by outcome. All learners should be aiming to 'meet expectations' for the final project. Some may exceed expectations and a small few may be categorised as 'working towards'.

Scaffolding
Activities have been appropriately scaffolded using a variety of techniques. Examples include step by step instructions, worked examples of code and partially created projects to help learners get started.

The guidance column in the lesson plan also provides specific, tailored examples for how to scaffold learning where necessary.

Build on this
Every learner should be aiming to 'Meet expectations' and complete the activities to a good standard. Scaffolding is provided to support this. There are some learners that will have a desire to go above and beyond and exceed expectations. Every lesson has a 'build on this' section that includes guidance on how learners can stretch their learning further.

Hardware and software guidance

EduBlocks
EduBlocks is a programming tool that allows users to learn Python by dragging and dropping blocks. EduBlocks allows students to build on the skills they learnt from programming using Scratch.

EduBlocks can be accessed via (Website IntR8).

Full information about setting up accounts and classrooms can be found on the EduBlocks website: (Website IntR9). This includes instructions for teachers to set up classrooms and instructions for students to join classrooms.

EduBlocks also provide 123 Coding Cards, which map Scratch to EduBlocks to Python: (Website IntR10). These can be printed as classroom resources or accessed online.

Scratch
Scratch is a free graphical block-based programming language developed for young people. Scratch can be used online through a web browser on a laptop, desktop computer or tablet. Alternatively, Scratch can be used without an internet connection by downloading and installing the offline version on a laptop or desktop computer. It is important to decide which option is best for you and practise using that method before a lessons with students.

Scratch can be accessed via (Website IntR11).

Full information about Scratch accounts can be found on their website (Website IntR12). This also links to specific information about teacher accounts and using Scratch offline.

Machine Learning for Kids
Machine Learning for Kids is a free tool that adds machine learning capabilities to the Scratch programming environment.
Machine Learning for Kids is an online website and tool available at (Website IntR13) and can be accessed from a web browser. An internet connection is required to use the website. No installation is required to use Machine Learning for Kids; however, we recommend trying out the lessons before using it within a class, in case you have any problems accessing the website.

Machine Learning for Kids has two major parts:

1. A web-based tool where you can set up a machine learning model and work with training data.
2. A version of Scratch with extension blocks for working with a machine learning model.

You can optionally create accounts for students on Machine Learning for Kids. Accounts are not required for the lessons in Chapter 1 (the only chapter in Stage 7 to use Machine Learning for Kids). Students can complete the whole chapter using the 'Try it now' feature. Using accounts offers additional features, including saving machine learning projects with training data.

If you do decide to create accounts for your students, you should follow any permissions processes required by your school and country.

Note that Machine Learning for Kids uses a customised version of Scratch. In Chapter 1, students will create a personalised project in regular Scratch so that they can save their work. They will then load their project into the Machine Learning for Kids version of Scratch so that they can train an AI tool to play their game or simulation.

You can find more information about Machine Learning for Kids, including how training data is stored, at (Website IntR14).

Note that Chapter 1 only uses numerical data. No personal information from students is used in training the machine learning models.

Online word processor with collaborative capability

We have used Google Docs as an example tool during Chapter 2 – Content creation, but you may use a different tool already within your school. The key feature is that it must allow real-time collaborative working from students at different devices and enable them to both write comments and track changes.

Google Docs can be accessed via (Website IntR15).

Students must have a way to share created documents with others, either with a shared storage area or by using the URL.

Full support for using Google Docs can be found at (Website IntR16). Select 'Google Docs Editors', and then 'Google Docs'.

Spreadsheet software

The Spreadsheet software used in Chapter 7 – The power of data is Google Sheets, but you can use any spreadsheet software that enables you to add conditional formatting. One advantage of using Google Sheets is that you can use Google Forms when capturing data and have it automatically populate a spreadsheet. If you choose other software this may not link as seamlessly.

Google sheets can be accessed via (Website IntR17).

Full support for using Google Sheets can be found at (Website IntR18). Select 'Google Docs Editors', and then 'Google Sheets'.

Data capture form software

Google Forms is used as the data capture form software in Chapter 7 – The Power of data. You can use any online forms software, but ideally it should be able to automatically feed your spreadsheet software.

Google Forms can be accessed via (Website IntR19).

Full support for using Google Forms can be found at (Website IntR20). Select 'Google Docs Editors', and then 'Google Forms'.

Micro:bit

A micro:bit is a programmable device used in physical computing.

Students will use the EduBlocks environment to create Python code for their micro:bits. They will test the code using the simulator in EduBlocks before connecting their micro:bit to the computer and downloading the code to the micro:bit. Once the code is on the micro:bit, students can use it connected to their computer or they can remove it and power it from a portable battery pack.

There are two versions of the micro:bit. The newer version has a processor, microphone, 25 programmable LED lights, 2 input buttons, a touch logo, GPIO pins, a speaker, a radio antenna, a compass, an accelerometer and a battery socket. The newer version is used in these books, although the original version can be used without sound options or alongside small electronic components to add sound.

A getting started video and text-based guide can be found on the micro:bit website (Website IntR21). There are also short CPD courses available for teachers at (Website IntR22).

Full features are shown via (Website IntR23).

Video recording equipment and storage

In Chapter 2 – Content creation, users will need to use video recording equipment. You will need to make sure that you have the relevant cables, SD cards or other storage devices. The chapter does not make any assumptions about what type of equipment or storage you have, so you can use the equipment you would normally use at school.

Colour printer

During Chapter 6 – Connect the world, you will need access to a colour printer in order to print students' work.

Chapter 1 – Our digital world

Project: Train a machine learning model to work with a game designed in Scratch

Chapter overview

In this chapter, students will recall elements of online safety from earlier stages in the course. They will practise responding to different scenarios that may occur when being part of online communities. They will then train a machine learning model to work with a game designed in Scratch.

Chapter 1: Lesson summary		
Lesson	**Learning Objectives**	**Summary**
1	• Identify ways to responsibly use technology safely, and how to protect wellbeing • Recognise the positive and negative consequences of online activity within a digital footprint	Students will recall how to stay safe online with additional considerations for their age group including reporting offensive or illegal behaviour. They will work through scenarios that they may encounter and discuss appropriate actions.
2	• Recognise that people can purposefully antagonise other online users and state the risks of engaging with them • List the benefits and risks of using online communities	Students will consider the challenges, benefits and risks of participating in online communities and develop ground rules for effective participation.
3	• Describe a range of applications of AI, including image recognition and in computer games	Students will describe a range of applications of AI including those that they have interacted with. At the end of the lesson they will play a game in Scratch where they have to stop messages from reaching an online troll. They will consider the inputs they are using when they learn to play the game.
4	• Describe how AI can take information from surroundings and produce outputs based on their models • Train a machine learning model to play a game in Scratch	Students will explore how AI can take information from the surroundings. They will learn that machine learning is a type of AI that is in common use. They will train a machine learning model to play the game that they played in the previous lesson.
5	• Personalise a machine learning project	Students will create a personalised version of the machine learning project provided. They will train a machine learning model to play the game and record the results.
6	• Share findings on a machine learning project	Students will share their findings on machine learning. They will consolidate their knowledge of how AI can take information from surroundings and produce outputs based on their models.

End of chapter project	**Train a machine learning model to work with a game designed in Scratch**
Example ideas that students could design:	Students will personalise a machine learning model that launches an object to intersect with a projectile. They will train a machine learning model to play the game and report on their findings. Examples of projects that the students could make include: • stopping messages being sent to an internet troll (example project) • preventing an asteroid reaching a planet • blocking a shot at a goal in a sport • stopping recyclable items being put into a regular waste bin (trash can) • preventing a harmful insect from reaching a plant.

Why we are learning this	
Useful skills developed in this chapter	• Machine learning is good for problems that have lots of good quality data and where the patterns in the data are hard for humans to spot. Some problems have so much data that traditional programming would take too long or be impossible.
Links to future careers	• Training machine learning and AI systems • Roles that don't yet exist in this emerging area
Links to other subject	• PSHE / Citizenship

Chapter 1 – Our digital world

Project: Design a robot to help someone who works at your school

Name	

	Working towards (1)	**Meets expected (2)**	**Exceeding expected (3)**	**Score**
Trained a machine learning game	• The Do not Feed the Troll game was trained with significant assistance	• The Do not Feed the Troll game was trained with minimal assistance. Findings were explained	• The Do not Feed the Troll was trained and findings were clearly explained	
Personalised a machine learning game	• The Do not Feed the Troll game was partially personalised	• The Do not Feed the Troll game was personalised with different costumes and/or sizes.	• The Do not Feed the Troll game was personalised with different costumes and/or sizes and additional code which demonstrates additional understanding of how the game functions	
Explain how the game uses machine learning	• Partial explanation of how the machine learning model learns from data	• Explains that the AI (machine learning model) learns to play the game from training data	• Explains additional details such as the relationship between the x position and power variables	

Teacher feedback	
Learner response	

Chapter 1:	Lesson 1 Safe and healthy use of technology		
Student's Book page 2			
Learning Objective/s	• Identify ways to responsibly use technology safely, and how to protect wellbeing • Recognise the positive and negative consequences of online activity within a digital footprint		
New keywords	**Cookie** – Small file of information that a website stores on your computer **Cyberbullying** – Bullying someone using online tools **Permission** – When a website or app asks to access your data or sensors like camera or microphone **Policy** – Document explaining what you can expect from a website or service, including how it will use your data **Wellbeing** – The way people feel about themselves and their lives		
You will need	• Make sure you are prepared to talk about the laws in your country relating to illegal online activity (such as cyberbullying and child exploitation). • Also make sure you can give advice on which authorities to report illegal online activity to.	**Key pedagogy**	Classroom talk

Activity	Instruction	Guidance

1	• Introduce the chapter. Explain that the first two lessons cover staying safe online and then you will move on to exploring applications of artificial intelligence.
	• Remind students of the definition of a digital footprint and the activities that can add to the trail created by your online activity (Page 2).
	• Introduce the discussion on the positive and negative consequences of a digital footprint.
		• Your students may have a wide variety of online experiences depending on their access to technology and the amount of freedom they are given.
		• Make sure the discussion is balanced.
		• Positive consequences include:
		  • Getting positive feedback on your work
		  • Interacting with people with shared interests
		  • Having a portfolio of your work available online when you apply for a job
		• Negative consequences include:
		  • Getting unkind feedback
		  • Damage to your reputation if you don't behave well
		  • Privacy concerns from third party access to your data

2	• Introduce the definition of wellbeing, and tips for protecting your wellbeing online (Page 3 and Page 1) Task A, 'Protecting your wellbeing online'. • Walk around the classroom and select examples to highlight to the class.	• Be aware that young people may have sensitive information that they do not want to share with the class. • Young people may be using online tools that they are not old enough to use. Remind them of age ratings, but make sure they are still prepared to protect their wellbeing.
3	• Explain that students have ways of taking action when they encounter worrying or inappropriate material online (Pages 3 and 4).	• Explain what behaviours are illegal in your country and how you can report them to the appropriate authorities. • Be aware that discussing these issues may cause some students to worry about whether they should report things they have seen online. Identify trustworthy adults with whom students can raise concerns.

Activity	Instruction	Guidance
4	• Present the material on personal responsibility for staying safe online (Page 4). • Demonstrate creating a project on the Scratch website and adding an 'Events', 'When Loudness >' block. If this is the first time you have done this you will see a permissions pop-up. Otherwise, you can click on the settings to see the permissions. Discuss why the website needs to access your permissions. Explain that the Scratch website is trustworthy and has policies in place to protect your data. • Introduce (Page 1) Task B, 'Stop and think!'. • Discuss the answers that students provide.	• Walk around the classroom and make students aware of any discussion topics that are relevant for the whole class. • Some students may have very strong views on protecting their data and not being tracked. There is no one correct answer on what you should do – the important message is to stop and think and to make a considered decision. • The Scratch website will ask for permission if you use features that require access to the microphone or camera. Scratch is a trustworthy website and it has a clear reason for requesting permission. There are also clear policies on the Scratch website.
5	• Introduce the practice scenarios in (Page 2) Task C, 'Practice scenarios'. • Go through the scenarios one at a time and discuss issues that arise. Students should make notes in their Workbooks.	• In the first scenario, Farah should tell a parent or guardian that someone has asked her to meet up. This will help to protect her wellbeing. • This is an opportunity to discuss any online safety incidents or concerns that are relevant to your school community, or events that are taking place in the world at the moment. • Relate the scenarios back to the concepts of wellbeing, personal responsibility and taking action. • Emphasise talking to a trusted adult if students are not sure. • While reporting is optional for inappropriate content, it helps to make the internet a safer and nicer place for everyone.
6	• Reflection: In pairs or small groups, ask students 'How will you act differently online after this lesson?' and then complete (Page 2) 'Reflection'. • Recap the 'stop and think' message.	• Remind students of any resources available to them locally if they are concerned about anything they encounter online.

Activity	Instruction	Guidance

	Homework (Optional) (Page 2)
	• Students are asked to discuss today's lesson with a trusted adult.
	  o Do they have any techniques they use for looking after their wellbeing when using online tools?
	  o Have they ever had to take action because of cyberbullying?
		• Encourage students to discuss the lesson with a trusted adult either at home or at school.
		• They should share their new learning about wellbeing when using online tools and prompt the adult for other useful techniques.
		• You could ask students to share reflections on their discussions at the start of the next lesson during the 'What do we already know?' bullets.

Build on this: **Make commitments**

Get students to write down their commitments to act differently online to protect their wellbeing, to take action when needed and to take personal responsibility for their data and permissions.

Chapter 1: Lesson 2 Responsible participation in online communities

Student's Book page 6

Learning Objective/s	• Recognise that people can purposefully antagonise other online users and state the risks of engaging with them • List the benefits and risks of using online communities		
New keywords	**Community guidelines** – Behaviours that are expected from members of an online community **Trolling** – Deliberately unhelpful behaviour in an online community		
You will need	• Check that you can access the Scratch website and know how to trigger the request for permission to access the microphone on the web browser you are using. • Option: print copies of the (Website C1L2R1), or similar guidelines from an age-appropriate community to use as a reference for writing online community ground rules.	**Key pedagogy**	Classroom talk

Activity	Instruction	Guidance
1	• Recap the definition of online communities and ask 'Which online communities do you know about?' (Page 6). • Age-appropriate online communities may include those related to games or toys for their age group. Students may also be aware of online communities in which their families participate. • Explain that Scratch provides an age-appropriate online community for children with an interest in creative computer programming. It's a great place to get experience of being a positive member of an online community.	• Make connections to online communities that are relevant to your students. • Do their parents or guardians use Facebook or Instagram? • Although students are not yet old enough to participate in some online communities it's important that they are prepared to.
2	• Recap the definition of trolling and effective ways to deal with trolls (Page 6). • Introduce (Page 3) Task A, 'Dealing with trolling behaviour'.	• While it might be tempting to try to reason with trolls, they are looking for a reaction and you are likely to encourage their behaviour. • Students are not responsible for other people's behaviour. They need to make sure that they behave well themselves and take appropriate action to protect themselves and others.
3	• Explain that there are risks and benefits to taking part in an online community (Page 7). • Ask 'How would you decide whether it makes sense to participate in an online community?'	• Students should consider whether a community is age appropriate. They could discuss their potential participation with an adult. They could find out whether the environment is positive or whether there is trolling behaviour. • Emphasise that they can always leave an online community at any time if they are having too many negative experiences.

Activity	Instruction	Guidance
4	• Discuss the importance of having guidelines for an online community (📖 Page 7). • Show students the Community Guidelines for Scratch (Website C1L2R2) • Put students into groups and ask them to complete (📖✏ Page 3) Task B, 'Online community guidelines'.	• If you have access to computers for this lesson, then you can share the link to the Scratch Community Guidelines as a reference. Otherwise, you could keep them displayed on the teacher computer as a reference. • Draw attention to the 'Embrace remix culture' in the guidelines – this is specific to Scratch. Would there be specific rules for your community?
5	• Reflection: In pairs or small groups, ask students 'What guidelines did you agree?' and 'How difficult would they be to follow?' and then complete (📖✏ Page 4) 'Reflection'. • Ask each group to share one of their rules.	• Discuss any issues that arise. Did they have disagreements within the group? • Acknowledge that it can be difficult to get the guidelines right the first time. Sometimes communities update their guidelines when problems occur.
6	**Homework (Optional)** (📖✏ Page 4) • Provide students with the links and ask them to find and read through the community guidelines for the Raspberry Pi Foundation and Tynker.	• These are the links to the pages: (Website C1L2R3) (Website C1L2R4) • You could start the next lesson by asking if they managed to find the guidelines and if so how.

Build on this: **More community guidelines**

Find out which age-appropriate communities your students participate in. Get them to find their rules or community guidelines, and discuss them. Did they know about the guidelines? Have they been following them?

Chapter 1: Lesson 3 Applications of AI

Student's Book page 8

Learning Objective/s	• Describe a range of applications of AI, including image recognition and in computer games
New keywords	**Artificial general intelligence** – Future computer systems that might be able to perform any task in a human-like way **Image recognition** – Identification of objects in images by computer systems using AI **Recommendation system** – Computer system that makes predictions on content that a user would like, using AI **Voice assistant** – AI system that can perform useful actions in response to voice commands

You will need	• Student computers that can access the Scratch website • A way to share a link to a Scratch project. (You can put it on the computer screen and have students type it if you don't have a better option.) The project students will need is (Website C1L3R1)	**Key pedagogy**	Classroom talk

Activity	Instruction	Guidance
1	• Recap what artificial intelligence is and cover the definition of artificial general intelligence (Page 8). • Lead the discussion on the topic of AI vs artificial general intelligence. • Introduce (Page 5) Task A, 'AI definitions'.	• Some students may be more familiar with the concept of artificial general intelligence (AI that can tackle any task in a human-like way). • Students may be familiar with artificial general intelligence from science fiction films or novels. • This could also be an opportunity to connect with stories about AI that are currently in the news. • Make sure that students understand the difference between AI as a practical subset of computing that is in use in many everyday technologies, and the theoretical future concept of an AI that could tackle any problem.
2	• Cover the material on applications of AI (Pages 9 and 10) including image recognition and computer games. • (Page 5) Task B, 'AI in your life'. • Ask some individual students to share an example. Also, share any key points that you have observed with the whole class.	• Observe students as they complete the task and make notes of any points to raise with the whole class. • Popular apps that use AI features include Duolingo (language learning), Roblox (computer game platform) and Netflix (video streaming service). • Students may find that they don't know whether a feature uses AI or not – for example, adaptive difficulty in an app for learning a foreign language. Mention that you can't always tell and they will be learning more in this chapter to help them to recognise AI systems.

Activity	Instruction	Guidance
3	• Explain that, in the next lesson, students will be training a machine learning model to play a game. To do this they will need to understand the data that is provided as input (Page 10). • Remind students that Scratch allows you to create your own apps. They will be playing a game that has been written in Scratch. • Introduce the 'Don't feed the troll' game. Explain that the player's job is to stop the messages reaching the troll (because reacting to a troll just encourages their negative behaviour). • Demonstrate loading the game but don't provide any information about it as students will be investigating. Share the link (Website C1L3R2) to the project. • Introduce (Page 6) Task C, How do you learn to play a computer game?' and demonstrate the game.	• Reassure students that it's fine if they are not good at playing the game! • Students are not modifying the game and do not need to remix or save it. • Students might be able to work out that the space key launches the stop sign but you could give some guidance to get them to reach this conclusion if needed.
4	• Reflection: In pairs or small groups, ask students to discuss 'How did you learn to play the game?' and then complete (Page 6) 'Reflection'.	• They probably used trial and error initially (they gathered some data). • Then they used the *x* (left to right) position of the target and their previous experience to predict how much power they would need.

5	**Homework (Optional)** (Page 6)
	• Students are asked to play the role of a machine learning plant identification app. They should look at plants they pass on the way home.
	    ○ What similarities and differences do they have?
	    ○ Do you get quicker at identifying plants when you see them often?
		• Students should think about how shape, size and colour help them to recognise plants.
		• You could start the next lesson by discussing the plants the students found. Did anyone find only one example of a certain plant? What did the most common plants look like?

Build on this: **Research uses of AI**

Students may not be sure if the apps they use make use of AI. They can research this to find out if popular websites, applications, games and other computer systems do make use of AI technologies.

Chapter 1:	Lesson 4 Train a machine learning model		
Student's Book page 11			
Learning Objective/s	• Describe how AI can take information from surroundings and produce outputs based on their models • Train a machine learning model to play a game in Scratch		
New keywords	**Machine learning** – Type of AI that uses data and algorithms to gradually produce more accurate results in a way that imitates how humans learn **Machine learning model** – Algorithm plus data that can be used to make decisions or predictions **Supervised machine learning** – Type of machine learning where humans label data to help the machine learning model to classify new data **Training data** – Data used to train a machine learning model **Unsupervised machine learning** – Type of machine learning where the AI detects patterns in data and learns how to classify new data		
You will need	• This lesson uses the Machine Learning for Kids website (C1L4R1). Make sure that you can access the website and complete the lesson from your school computers. • Work through the project yourself before the lesson, so that you will be able to provide assistance. • Save the starter project (The 'Do not feed the troll' starter project https://scratch.mit.edu/projects/923711779/) to a place on your computer network where students will be able to open it in Scratch.	**Key pedagogy**	Develop practical skills

Activity	Instruction	Guidance
1	• Introduce the definitions of machine learning and material on training a machine learning model (Page 11). • Give students time to complete (Page 7) Task A, 'Complete the AI definitions' and (Page 7) Task B, 'Labelling data'.	• Make it clear that machine learning technology is used widely in chat bots, facial recognition systems and computer games. This is not futuristic artificial general intelligence. • Machine learning learns from examples. The training data needs to match the problem. • Make sure you leave enough time for the practical part of this lesson.

Activity	Instruction	Guidance
2	• Introduce the 'Train a machine learning model' section (📖 Page 12). • Tell students where they can get the starter project from. This will be the place where you have saved it on your school network. • Direct students to follow the instructions to train their machine learning models and to complete the Workbook activities as they go along (📖✏ Pages 7 and 8). • Explain that they will use a different version of Scratch. Show them how to turn on high contrast blocks so that their blocks look the same as those in the book.	• You could either save the starter project to your school network, or have the students download the project and then load it into Scratch. • You cannot use regular Scratch for this project – you need the machine learning blocks. • Make sure that students realise that they need to complete the Workbook activities as they go along. They will not be able to record the earlier data if they leave it until later. • Reassure students that it doesn't matter whether or not they are good at playing the game.
3	• Observe and troubleshoot as students complete the activity. • When you are near the end of the lesson, tell students that they need to record their final results by completing (📖✏ Page 8) Task E, 'After many attempts'.	• Make sure students have selected the correct sprite when adding code. • Check code very carefully. The blocks and all their inputs must match exactly. • Students must choose the 'lose' option in the lose script on the Target sprite. • The name of the 'recognise numbers ... (label)' block is very long. Make sure they choose the version with '(label)' at the end, to get the label that Machine Learning for Kids has categorised the data with. • The '*x* position of Target' block is provided on the code area to drag in. If they need to recreate it, then you need to change the dropdowns for the 'Sensing'> 'backdrop # of Stage' block. • Training data will not be saved unless students have created accounts. This is okay as you will not need this data for the next lesson.

Activity	Instruction	Guidance
4	• Tell students that they do not need to keep their training data or projects. Next week they will be personalising the game and training a new model. • Reflection: In pairs or small groups, ask students to discuss how the training data improved the AI and then complete (Page 8) 'Reflection'. • Mention how important it is to have lots of good-quality training data for training machine learning algorithms.	• Students should find that the machine learning model improves at playing the game with more data. • They may notice that the machine learning player is more likely to lose at the edge of the Stage. • They may notice that the machine learning player has close misses as they get better. • They will each get different results due to the different training data they have provided. • If they let the machine learning model play for long enough, then they should see more wins than loses in their training data.
	Homework (Optional) (Page 8) • Choose a project from zooniverse.org for students to explore. They don't need to sign up but can practise identifying images.	• This is an example of a project where volunteers identify elephants from distinguishing features <u>(Website C1L4R2)</u>.

Build on this: **Keep playing**

Let the machine learning model play a lot more times. Does it get really good? Do you get a lot more wins than loses?

Chapter 1: Lesson 5 Make your own machine learning project

Student's Book page 16

Learning Objective/s	• Personalise a machine learning project		
You will need	• <u>(Website C1L5R1)</u> • Work through the project yourself before the lesson so that you will be able to provide assistance. • The 'Don't feed the troll' starter project <u>(Website C1L4R2)</u>	**Key pedagogy**	Develop practical skills

Activity	Instruction	Guidance

1	• Explain that in this lesson, students will be creating their own personalised version of 'Don't feed the troll'.
	• Talk through the example projects.
	• Demonstrate loading the 'Don't feed the troll' starter project (Website C1L5R3)
	• Talk through the roles of the three sprites.
		• Make sure students are clear on the roles of the sprites and the kinds of project they can make. They will not be able to make major changes.
		• This lesson is another opportunity to use Machine Learning for Kids and understand the importance of training data.
		• Examples of projects that the students could make include:
		  ◦ stopping messages being sent to an internet troll (example project)
		  ◦ preventing an asteroid reaching a planet
		  ◦ blocking a shot at a goal in a sport
		  ◦ stopping recycling being put into a regular waste bin (trash can)
		  ◦ preventing a harmful bug from reaching a plant.

2	• Explain that students can personalise the project by changing the costumes and adding a backdrop. • Make some minor changes then demonstrate saving the project and closing Scratch. • They can also adjust the size of the Launcher and Projectile to control the difficulty. • Tell students to work through planning and personalising their projects (Pages 16 and 17). • Direct them to complete (Page 9) Task A, 'Plan your project'. • Emphasise that they should make sure that their projects are working before moving on to creating the machine learning model. • Students must save their projects to their computers as .sb3 files so that they can load them into the Machine Learning for Kids version of Scratch.	• Students should first create their projects in Scratch website. Once they begin training their machine learning models, they won't be able to reopen their projects unless they have an account at Machine Learning for Kids (not required). • Students must limit their modifications to those suggested to ensure that they will be able to train the machine learning model to play their game. • Put a time limit on personalisation to ensure that students have time to train their models.

Activity	Instruction	Guidance
3	• Observe students and provide assistance as needed. • When they are ready, move them on to creating and training their machine learning model (Pages 17 and 18) and recording their results in their Workbook. • Direct them to complete (Page 10) Task B, 'Record your results'.	• Refer students back to the detailed instructions in the previous lesson if they need further assistance. • Remind them to check the code very carefully. • Remember that the high contrast setting makes the blocks look like the ones in the Student's Book.
4	• Ensure that students have recorded the results of training their game. • Reflection: In pairs or small groups, ask students to discuss 'What would happen if you changed the game by moving the Target to a different position or changing the starting position of the Projectile?' and then complete (Page 10) 'Reflection'.	• If you change the game significantly, then the AI will not be able to play it well using the existing training data. It would need new training data to match the new environment.
	Homework (Optional) (Page 10) • Students are asked to discuss with a trusted adult whether a computer could be taught to play any game.	• You could ask students to share insights at the start of the next lesson. • Ask students to discuss examples of games that would be easy and games that would be hard to train.

Build on this: **Change the data**

What happens if you change the game? Try changing the *y* position of the barrier. (Guidance: you would need another number data item for the *y* parameter, and you will need to paste this in when you get a prediction.)

Chapter 1:	Lesson 6 Share findings on AI and machine learning		
Student's Book page 20			
Learning Objective/s	• Share findings on a machine learning project		
You will need	• Workbooks with results from the previous two lessons.	**Key pedagogy**	Showcase

Activity	Instruction	Guidance

1	• Recap that students have:
	  • trained a machine learning model by playing the game and using random data
	  • used the training data to get a machine learning model to play the game
	  • continued adding training data from the machine learning model's attempts to play the game (Page 16).
		• Reiterate that machine learning technology is in common use as training for specific tasks, whereas artificial general intelligence is theoretical future technology that could solve a wide variety of tasks in a human-like manner.
		• Remind students that they are using the same technology that is used for many everyday AI computer systems.

2	• Put students into groups and get them to showcase their results to each other and then summarise their findings across the group.
	• Each student should record their shared findings in their Workbook. Ask them to complete (Page 11) Task A, 'Record your findings'.
		• Ask questions about how the training data relates to the performance of the AI and how the AI's performance changes over time.
		• Findings could include:
		  • The AI performed badly if it didn't have enough training data.
		  • The AI improved over time.
		  • Eventually the AI got very good and was mostly winning.
		  • Making the game harder (through smaller sprites) meant more training data was needed.
		  • When the AI only makes a few mistakes they are usually a near miss.
		  • The AI takes time to check whether it's power predictions are likely to win.
		  • Sometime the AI runs out of guesses – AI can use a lot of computing resource.
		  • If you change the game then the AI won't work well as it is training on different data.

Activity	Instruction	Guidance
3	• Go round the groups and ask for a different key point from each. Record these on a whiteboard. • Encourage students to add any additional findings to the list their Workbooks.	• Were there any surprising results? Did anyone try anything unusual like training the AI only on losing shots? • Did anyone try making the game really easy or really hard? • Did anyone's AI fail to learn (did they correctly label the data with the 'win' and 'lose' dropdowns?) • Remember that there will be some variation in results due to the different training data and the personalisation of games.
4	• Ask 'How else could you have programmed a computer player for your game? Which would you prefer?' (Page 20). • Cover the material on machine learning vs traditional programming.	• For this example, they could have used maths to calculate the intersection point. • Some of your students may be confident mathematicians who would love to use maths to solve the problem, Others would much prefer to play a game to create the training data. The appropriate solution for a particular problem does partially depend on the people who are available and their skills and interests.
5	• Ask students to complete (Pages 12 and 13) Task B, 'Reflection'.	• Students can refer to their Student's Books and Workbooks if they need additional support.

Activity	Instruction	Guidance

6	• Reflection: In pairs or small groups, ask students to discuss 'What kinds of problems is machine learning good for?' and then complete (Page 12) 'Reflection'.
	• Students can then complete the (Page 12) Project skills checklist.
		• Machine learning is good for problems with these characteristics:
		  • Lots of complex data
		  • Hard for humans to spot patterns
		  • Training data available
		  • Model can be reused once trained
		• If students give specific examples, such as a voice assistant, ask them why machine learning is good for that application. (It would be very hard for a human to use traditional programming to turn a sound wave into words. A machine learning model can use its algorithm and lots of training data to recognise the words.)
		• Machine learning uses a lot of computing resources so it should be used for high value problems, and problems where the model can be reused once it is trained.
		• There also needs to be a way to collect training data in an environment similar to the conditions in which the AI needs to operate. The AI trained for one game wouldn't perform well in a different game.

	Homework (Optional) (Page 12) • This asks students to identify three problems that would be best solved by machine learning.	• You could leave this question open or give students categories: for example, one problem from space, one from land and one from water.

Build on this: **Hold a debate on machine learning vs traditional programming**

Hold a debate on whether it's more beneficial to train a machine learning model or write traditional code to solve specific problems. Examples could include analysing medical symptoms or writing a quick program to calculate the cost of party supplies.

Chapter 2 – Content creation

Project: Design and record a news report that makes a prediction about future technologies

Chapter overview

In this chapter, students will work collaboratively using online tools. They will investigate the reliability of sources of online news coverage and futurist predictions. Students will build on their knowledge of plagiarism, crediting, citation and fair use to correctly use and attribute images. They will work in groups to create a collaborative script and record a video of a future technologies news feature where they each play a different role. Students will think about storage capacity when saving their videos to memory cards and transferring their videos from their computers to the school network.

Chapter 2: Lesson summary		
Lesson	**Learning Objectives**	**Summary**
1	• Practise typing skills • Use comments and track changes in collaborative documents	Students will practise their typing skills using an online collaborative word processor while they work together to create a news article. They will use their experience to create guidelines for collaborative writing, and reflect on their shared experiences.
2	• Describe the advantages and disadvantages of instant access to news and current events • Use advanced search techniques in a search engine to refine results • Describe plagiarism, crediting, citation and fair use	Students will investigate the advantages and disadvantages of real-time access to global events through the internet. They will explore the reliability of information by looking at sources and citations. Students will practise searching for images, understand their licences and the permission requirements for these use.
3	• Explore future technologies and make predictions about them	Students will look at how experts use the internet to share future technology predictions. They will look at example predictions from the past and search for current trends. Students will understand why trust is important and analyse their current trend findings to judge how trustworthy they are.
4	• Use online and offline tools to design and collaborate	Students will be introduced to the project brief and work in groups to create a script for a future technologies news report. They will assign roles and build the script collaboratively.

5	• Use devices to create digital artefacts using video • Describe remote file storage options including on other computers, across networks or in the cloud • Identify storage capacity limitations across a range of devices	Students will explore video files with a focus on storage implications of video. They will plan and rehearse their news reports before recording them as a single take. Students will save their best take to the school network and link to it in their script.
6	• Showcase news reports	Students will take part in a news broadcast lesson where they each share their final news report. They will discuss reports as a class to evaluate how realistic each future technology prediction is.

End of chapter project	**Design and record a news report that makes a prediction about future technologies**
Example ideas that students could design:	The students will record a news report that makes a prediction about a future technology. The students will give themselves roles such as presenter, interviewer, expert and cameraperson. The following examples of possible future technology predictions have been included in the book for inspiration: • a flying car • home 3D printed shoes • a lab grown dodo bird.

Why we are learning this	
Useful skills covered in this chapter	• Writing scripts and recording video
Links to careers	• A variety of careers such as TV news journalist, documentary writer or presenter, vlogger
Links to other subjects	• Script-writing skills in English

Chapter 2 – Content creation

Project: Design and record a news report that makes a prediction about future technologies

Name	

	Working towards (1)	Meets expected (2)	Exceeding expected (3)	Score
Video creation	• A video recording that either doesn't match the script or is not centrally stored	• A video recording of the news report transferred from the device to central storage	• A video recording of the news report transferred from the device to central storage. Improvements have been taken into account	
Collaboratively created script	• A script document with only partial scripted responses or no labelling of roles	• A script document with labelled roles and interview answers in full or as bulleted points	• A script document with roles, camera directions, and props labelled. Interview answers are written in full or as bulleted points	
A trustworthy future technology prediction	• There is no clear evidence to suggest why the subject of the news report has been chosen	• The subject of the news report has been chosen based on research	• The subject of the news report has been chosen based on research and trust has been established	

Teacher feedback	
Learner response	

Chapter 2:	Lesson 1 Collaborative working		
Student's Book page 23			
Learning Objective/s	• Practise typing skills • Use comments and track changes in collaborative documents		
New keywords	**Online collaboration** – People working together in apps accessed over the internet **Coworking buildings** – Workspaces where people go to work together with others in the same physical location		
You will need	• An online word processor with collaborative working enabled, such as Google Docs (Website C2L1R1)	**Key pedagogy**	Working collaboratively

Activity	Instruction	Guidance

1	• Go through the 'What do we already know?' bullet points with students (📖 Page 23).	• Students will be very familiar with the term word processor as they have used word processors since Stage 1. They may not have used an online word processor and, even if they have, they may not have used one in a collaborative way.
		• If students followed this course in earlier stages:
		  • In Stage 3 Create with code, they listed the benefits of working with others when programming. Much of this knowledge and experience will be valuable when working collaboratively in a word processor during the current chapter.
		  • In Stage 4 Our digital world, students described how digital communication enabled online communities to exist and changed the way people interact. During that lesson they also demonstrated how to be respectful in an online community. This background knowledge will help students to understand how online collaboration is possible, the benefits of it, and the behaviour that is appropriate.

Activity	Instruction	Guidance

2	• Read through the online collaboration tools content, highlighting the use and types of tools (Page 23).
	• Ask 'When have you used online collaboration tools? Why did you use them?' and then lead into the advantages of online collaboration tools.
	• Talk through the example showing co-authoring of a document (Page 24).
	• Tell students about the commenting and track changes tools, referencing the images in the Student's Book.
		• Students are likely to have used or be familiar with online collaboration tools. Features not specifically mentioned in the Student's Book that might be relevant are:
		  ◦ collaborative file sharing such as shared photo albums or drives
		  ◦ collaborative communications tools such as Zoom, Discord or Microsoft Teams.
		• Make reference to any online collaborative tools that students may have already used in school.
		• When talking through the illustration showing three people working collaboratively on a document, you could highlight that the people are in different locations and using different types of device. They are all in the same document and you can tell this by their avatars in the top menu bar. You can also see which position each person is at by their coloured marker in the page content.

Activity	Instruction	Guidance

3	• Demonstrate opening your chosen online word processor, then start a new document. Demonstrate:
	  ◦ creating a new document
	  ◦ sharing a document with others
	  ◦ seeing who is online
	  ◦ seeing where users are in the document
	  ◦ adding text
	  ◦ someone else editing the text
	  ◦ both adding text
	  ◦ leaving a comment and/or a suggestion instead of direct editing
	  ◦ resolving a comment
	  ◦ how to track changes to see what has been edited.
	• Draw attention to the Stay safe! tip (Page 24).
		• As you demonstrate using your chosen tools, you should point out the same things that were shown in the illustration of three users collaborating in a document. For example, your avatar/initials and your coloured marker.
		• When accessing the document from a different machine, you can ask another member of staff or a student to do this with you. Your co-author doesn't need to be in the same room but must be well prepared in advance so that the focus is kept on the main parts of the demonstration.
		• Demonstrate the safety features. For example, when sharing with someone else, show the options such as 'anyone can access' / 'anyone with a link can access' and explain that you want to control who has access and the level they have ('editor', 'commenter'). These settings will differ depending on your chosen tool.

4	• Organise students into pairs or small groups. They should sit next to each other but use their own computers. • Get students to log on to the computers and access the online collaborative word processor you have chosen. • Tell students that one person should create the document and share it with the others. • Ask students to complete (📖✏ Page 13) Task A, 'Breaking news!'	• Make sure that all groups have created and shared their documents successfully. They should see all members of the group in the document before they begin typing. Some students might have an impulse to start writing as soon as they have access but this should be discouraged as it will create catch-up pressure for others, which may make them feel excluded. • A focus of this lesson is practising typing. Students have been improving their typing skills throughout all stages. Remind them about placing their hands correctly on the keyboard using specific tips local to you, such as the homing bars on the 'F' and 'J' keys of most QWERTY keyboards. This is also a good opportunity to ensure that students are positioned correctly and sitting properly to avoid repetitive strain injury. • Keep group size small to get the best experience of collaborative working. You could explain that you are deliberately keeping the group small because large groups working collaboratively often have problems when trying to edit the same parts, and someone may become excluded. The students are likely to experience this to some extent as they start to work collaboratively in this way, but it will be more controlled. • Observe students and encourage everyone to participate. Also make sure that students are participating from their own machines. They should write, edit and format the article to achieve the best result. • At the end of the activity you could ask students if they want to share their headlines. Students may have taken completely different

Activity	Instruction	Guidance
		approaches to the content/viewpoint of their articles.

5	• Ask students to complete (Page 14) Task B, 'Co-author reflection'.
	• Ask 'What are your best ground rules for successful collaborative working in a word processor?' (Page 25). Discuss the ground rules and compile a class list on the board.
		• Students should all have found some advantages and disadvantages of collaborative working. In some cases this will be because the tools and this method of working are new to the students. Others will be ongoing considerations.
		• Ground rules could include:
		  • Don't edit the same sentence or section as someone else.
		  • Use comments or suggestions when editing.
		  • Agree clear roles and responsibilities.
		  • Look out for other people and help them to collaborative if they are feeling excluded.
		  • Use comments to add praise when work is really good.
		• Keep a note of the class list of ground rules as this will be useful in Lesson 4.

6	• Consider students' experiences by having a class discussion based on what they have written in their Workbooks as part of the co-author reflection task.	• Ask students to imagine a scenario where they always have a computer and access to the internet so that they can work collaboratively from anywhere. • They may choose to always work this way because it's more efficient, they do not need to leave their house and travel to school, and they could collaborate more often by using tools that make it easy. • Alternatively, they may say that they would miss the face-to-face interaction, they would worry if they're not as quick using the tools, or they don't like seeing their work being edited or commented on by peers in this way. • Highlight that balance between the two approaches is a good option.

Activity	Instruction	Guidance
7	• Reflection: In pairs or small groups, ask students to discuss 'If all of your collaborative work could be completed online from anywhere with a computer, why would/wouldn't you want to do it that way all of the time?' and then complete (Page 14) 'Reflection'.	• They may think of positive things, like being able to travel and work at the same time or being at home to be around for caring responsibilities or saving money on travelling for work. • Negative things might be that they feel there would be less interaction with wider groups of people.
	Homework (Optional) (Page 14) • Students are asked to watch the local TV news or listen to a news report on the local radio. Then they write down a suitable headline for the news report and the key facts that were presented in the report.	• You could also suggest that they read a local newspaper.

Build on this: Instead of sitting next to each other when collaborating, you could have students sitting across the room or in another room to see what additional challenges this brings. They could discuss ways to reduce the impact, for example, use the chat facility to communicate, or use an online video call for instant discussion.

Chapter 2: Lesson 2 Usage rights

Student's Book page 26

Learning Objective/s	• Describe the advantages and disadvantages of instant access to news and current events • Use advanced search techniques in a search engine to improve results • Describe plagiarism, crediting, citation and fair use		
New keywords	**Plagiarism** – Copying someone else's work then saying it is your own work **Citation** – Credit given to identify the original author of a piece of work **Fair use** – Exceptions that allow the use of work that is under copyright		
You will need	• High level information about local copyright and fair use laws (optional) • Your shared word processor tool if it has a plagiarism checker (optional) For example Google Docs	**Key pedagogy**	Real-world contexts

Activity	Instruction	Guidance
1	• Read through information about the platforms for news coverage (Page 26). • Ask 'Why might someone feel excluded from what is happening in the world today?' and 'How could an increase in the availability of online live news coverage have a negative effect on someone's mood?' (Page 27). Students could think about these questions in pairs before sharing their thoughts with the class.	• You could demonstrate to students how you can access real-time global news during class by pulling up a news website or engaging with a smart device. • When discussing how someone could feel more excluded, encourage students to think about the barriers to accessing news from these platforms. Maybe they do not have smart devices or internet access. Perhaps they are too young to access social media. Accessibility also plays a part, as content that is created quickly without sophisticated editing might not be as inclusive for those with disabilities. • When discussing how availability can negatively impact mood, encourage students to think about which type of news coverage makes the most headlines. Often main news items included stories of war, natural disasters and disputes. With these stories being available almost continuously, people might feel sad, angry and powerless. With news companies in competition to break the stories first, the footage can be unscripted and graphic, which can cause distress for people who consume the footage and anger at a lack of privacy for the subjects of the news.

2	• Talk through the six 'online live news consumption' images (Page 27) using additional information from the Teacher Guide guidance notes.	• Highlight the convenience of news consumption:
		  ◦ consumers see the news embedded into their existing apps
		  ◦ the sophistication of social media algorithms means that the news shown is based on interests of the consumer and popularity in their demographic. There is a greater chance of seeing news that is relevant to the user.
		It is worth pointing out, however, that, with the speed at which social media feeds update, consumers may miss relevant news. Also point out that social media platforms are built for entertainment and socialising, rather than for news sharing.

Activity	Instruction	Guidance
		• If possible, refer back to the Stage 6 'Our Digital World' work on digital footprints. It is possible to see a track record of a news source through their digital footprint. However, it is worth pointing out that not everyone will know how to check a digital footprint and so will not know who can be trusted. They may mistakenly trust someone with a bad track record for accurate reporting. • Mention that with more news sources available many consumers have a greater choice of where they actively get their news from. Some sources may be politically or commercially influenced while others may not. As with the second image, it's worth noting that influence is not always obvious. This means bias may be fairly well hidden but will ultimately not reflect a balanced view of reality. • Tell students that some news sources offer paid subscriptions to access their news. Others offer their content for free but users need an account. This can be a good way for consumers to access news, but paywall or registration gates can act as a barrier, as not all sources act responsibly with user data. Online subscription costs will exclude some people from accessing high-quality news coverage. • One positive aspect of instant access to local breaking news is that if help is needed people will know about it quickly and can get there, or if an area is at risk people can stay away. A negative aspect is that live local news can put people at risk because their location is shown. Also some people will want to go towards the source of the story even if the advice is to stay away. • News that makes headlines can misrepresent the danger in an area and this can have an impact on tourism and the local economy. This can also happen with health news where misrepresentation leads to an unbalanced view of risk and can exacerbate a problem.

Activity	Instruction	Guidance
3	• Ask 'What examples have you seen where instant access to global news has been a good thing?' (Page 27).	• Encourage students to think about stories they might have seen globally. Perhaps they have seen stories where aid was required or missing people were found, or an uplifting story changed their mood. Additionally, they might have seen something locally, maybe when an event was cancelled at the last minute and it saved people a trip, or when a product needed to be recalled urgently.
4	• Describe plagiarism, citation and fair. Ask 'Why is citation important?' (Pages 28 and 29). • Highlight the Stay safe! tip as students may be concerned that they have accidentally plagiarised work in the past.	• You could highlight culturally relevant plagiarism court cases, for example, between musicians or authors. • If the online collaborative word processor you used last lesson has a plagiarism checking tool, then you could demonstrate this by typing a piece of your own text and also a piece of copied text. Explain that even your own text won't have a score of zero.
5	• Explain that search engines have advanced filtering tools, such as the ability to filter by usage type (Page 29). • Read the Stay safe! tip to highlight other advanced filtering techniques. • Demonstrate how to perform a usage rights filter in your chosen search engine.	• The specifics of your demonstration will depend on the search engine you use. In Google Images, there is a 'Tools' button that brings up a submenu. You can then show all images or restrict them by different licences. Search for an image, such as 'sunset', and then ask the students to watch the visible images change as you click on each licence type. Highlight images that have a 'Licensable' icon in the bottom left-hand corner and click on an example to see the licence details.
6	• Set up students on computers and remind them of your chosen search engine. • Ask them to open a document as they will be pasting their images into it to share with the class. • Direct students to complete (Pages 15 and 16) Task A, 'Image usage rights'. • Ask students to share their images with the class or in small groups. Did they choose the same images?	• See the licence type answers in the Answer document. • Other advanced search tools will vary by search engine, but for Google, they will be size, colour, type and time. • If students find images that they really want to use but can't, then they might feel frustrated or annoyed but they should understand the reasons why, and have sympathy with the owner who set it up that way.

Activity	Instruction	Guidance
7	• Reflection: In pairs or small groups, ask students to discuss 'Did you find images that you wanted to use but weren't allowed to? How did you feel about this?' and then complete (Page 16) 'Reflection'.	• This can be a frustrating scenario but it is important that students understand that they must only use images that they are allowed to use.
	Homework (Optional) (Page 16) • Students are asked to interview someone from an older generation. It could be an adult in their household or a member of their local community. They should ask them how they heard about the latest news when they were younger.	
Build on this: Instead of talking through the online live news consumption images you could use them as starting points for a class debate. Positives, negatives and other talking points are highlighted in the guidance column above.		

Chapter 2: Lesson 3 Future technologist reliability

Student's Book page 30			
Learning Objective/s	• Explore future technologies and make predictions about them		
New keywords	**Future technologist** – Expert who uses current technology trends to make predictions about the future		
You will need	• No extras	**Key pedagogy**	Real-world contexts

Activity	Instruction	Guidance
1	• Define the role of a future technologist (Page 30). • Ask 'Which platforms do you think a future technologist could use to reach their audience?'	• Future technologists could make use of any of the platforms discussed in the last lesson such as blogs, podcasts, videos and social media. They could also feature in traditional media such as print or scheduled programme. In addition, they might embed themselves into online communities around their prediction theme (online communities were discussed in Stage 4 and in Chapter 1 of this Stage). • Using targeted advertising provided by social media platforms which use algorithms to personalise people's advert experience.
2	• Use the examples (Page 30) to highlight that not all predictions become reality. • Set up students on computers, remind them of your chosen search engine and then complete (Page 17) Task A, 'Researching predictions'.	• In addition to writing down their findings you should encourage students to cite the source in their Workbooks and to bookmark the useful sites they found, as they will be referring to them later.
3	• Ask 'Why is it important to look into the trustworthiness of a future technologist?' and 'How can a future technologist influence decisions now?' (Page 31).	• Future technologists might have a digital footprint that shows they have a reputation for getting predictions wrong. Sometimes people can become popular because of the way they market themselves, rather than the accuracy of their content. • Future technologists have the power to influence people's spending by encouraging people to buy technology or shares in technology companies. They can influence the skills people choose to learn and careers they decide to enter into.

Activity	Instruction	Guidance
4	• Talk through the methods for establishing trust in online content. • Set up students on computers. • Ask students to discuss 'How trustworthy were the predictions you found?' and to complete (Page 18) Task B, 'Establish trust'.	• Students will need to recall the websites they cited and bookmarked in Task A. • Students can use any extra time to search for more trends and predictions or to search for other online content that backs up their predictions.
5	• Reflection: In pairs or small groups, ask students to 'Share the prediction that is most likely to happen based on the research you have carried out' and then complete (Page 19) 'Reflection'.	• Did all students agree on the same prediction? • Did some users discount a prediction that others felt was trustworthy? • A good example to add of a kitchen appliance that is now often internet enabled is an oven – many models are now internet enabled so that they can be remotely controlled. Many cars are internet enabled so that they can be started without the driver being in the car, and they can alert their owner if they have been started (stolen?) for example.
	Homework (Optional) (Page 19) • Students are asked to choose an item of technology that is familiar in their household, for example a microwave or a toaster. They then need to make a prediction about how this technology could change in the future.	

Build on this: If you have time, you can play videos of TV programmes from the past that predicted what technologies would exist now. There are some called 'Tomorrow's World' that can be found on YouTube or the BBC Archive site that show school children predicting the future. How many predictions came true? Discuss why students think that the children thought these predictions would come true. (Perhaps the news or science fiction influenced them.)

Chapter 2:	Lesson 4 Plan a news report		
Student's Book page 32			
Learning Objective/s	• Use online and offline tools to design and collaborate		
New keywords	**News script** – Text that details dialogue, timing and camera actions of a news item **Camera direction** – Shots and angles used to describe the position, view and movement of a camera		
You will need	• An online word processor with collaborative working enabled, such as Google Docs • A video of a local news report that is in interview style	**Key pedagogy**	Working collaboratively

Activity	Instruction	Guidance
1	• Introduce the project brief and talk through the images (Page 32). • Talk through the roles that students will play in their news videos. • Play a video of a news report in interview style. • Tell students that they will be planning their scripts in a collaborative word processor. Ask 'Why would collaborative tools be useful for creating a script?' and 'What ground rules aid collaborative working?'.	• Share a local news report that is an interview style video. • Ask students to pay attention to the roles of the people on screen and to the camera movements and angles. • If there is an additional role in your news report, you can add it to the list of roles students will have for their projects, and make the groups bigger accordingly. • Students should realise that having all of them inputting collaboratively into the script will make use of everyone's skills and knowledge, and will improve the end result. • After discussion, remind students of the class ground rules for collaborative working that they compiled in Lesson 1.

Activity	Instruction	Guidance
2	• Organise students into groups. • Ask them to complete (Page 20) Task A, 'News report overview'.	• Resolve any disagreements about roles by encouraging students to compromise for the good of the project. • When deciding on a topic for their report, students should first reflect on their findings from the last lesson and discuss them with their group.
3	• Talk through the important features of a news script (Page 33). • Use the illustration to point out how the speakers and camera movement are clearly labelled. • Ask students to complete (Pages 20 and 21) Task B, 'Interview observations'.	• Replay the localised news interview video while students complete the Workbook task. • Students should note that the presenter does an introduction at the beginning and a summary at the end. They should be able to recall how the script, camera and props were used. They should notice how each of the people refers to the others when talking directly to them or introducing them to the audience.

Activity	Instruction	Guidance

4	• Introduce the Build task and set up students on the computers.
	• Students should complete (Pages 21 and 22) Task C, 'News report notes' as they discuss and write their scripts.
		• Remind students that they created and shared a collaborative document in Lesson 1 of this chapter.
		• Students will collaborate offline through discussion and note taking and then online in the document. You could also encourage students to collaborate offline on a storyboard or mind map to help them to think creatively.
		• Think about how to position the groups.
		  ◦ Will they be at computers next to each other?
		  ◦ Will they be dispersed but have access to breakout meeting software?
		  ◦ Will they gather together to plan roles and responsibilities and then disperse to work on their documents, using comments to chat?
		• Remind students that if they use a source for their information, they should cite the source.
		• You could print copies of the final scripts ready for next lesson if students would like to read from them, or use them as props.

5	• Ask 'What were the advantages and disadvantages of working collaboratively on the script?' (Page 33). • Remind students to bring any props they need for recording their report to the next lesson.	• Students might find that because they have set roles in the news report it is easier to collaborate on this script than it was with the news article in Lesson 1. • Students might also reflect that it is easier now that they have experience of using collaborative working. • Alternatively, students might find that it is more difficult to write a linear script when they are dependent on another person's narrative before they can add their own.

Activity	Instruction	Guidance
6	• Reflection: In pairs or small groups, ask students to discuss 'What were the advantages and disadvantages of working collaboratively on the script?' and then complete (Page 22) 'Reflection'.	
	Homework (Optional) (Page 22) • Students are asked to find a local news article and practise reading it out loud to a pretend audience or a member of their household. This will help them to develop their presentation skills for the next lesson.	

Build on this: You could tie this project to a creative writing piece where students can choose any future technology prediction, no matter how unlikely. This will involve some more research using key words to ensure that they get imaginative, distant future predictions or predictions based solely on fiction.

<table>
<tr><th colspan="2">Chapter 2:</th><th colspan="3">Lesson 5 Record a news report</th></tr>
<tr><td colspan="5">Student's Book page 34</td></tr>
<tr><td>Learning Objective/s</td><td colspan="4">• Use devices to create digital artefacts using video
• Describe remote file storage options including on other computers, across networks or in the cloud
• Identify storage capacity limitations across a range of devices</td></tr>
<tr><td>New keywords</td><td colspan="4">Storage capacity – The maximum amount of data that can be stored
Cloud storage – Space to save, access and manage files that is accessed over the internet</td></tr>
<tr><td>You will need</td><td colspan="2">• Video recording equipment
• Internet enabled cameras, computer-to-camera data cables, or computers with SD card readers
• Props for recording (could be supplied by the students) and printed scripts if students have requested them
• Pre-recorded short videos of 30 seconds, 1 minute and 5 minutes that have been transferred to your computer so that you can demonstrate their file size (these are purely to demonstrate file size, not to play to the class)</td><td>Key pedagogy</td><td>Develop practical skills</td></tr>
</table>

	• A collection of different files that have different file types such as a word document, an audio file or an image file. The content of the file is not relevant to the task, it is purely for demonstration purposes to show students the different file sizes • Space for groups to record their videos		

Activity	Instruction	Guidance
1	• Explain that students will be using this lesson to record their news reports. • Talk through the types of video recording devices and explain that videos have large file sizes. Explain the different common options for storage and transfer (📖 Page 34). • Show students your pre-recorded short videos in a file manager, and highlight the differences in file size.	• As a comparison, you can show students the sizes of other files such as a text document, an audio file and an image file. • If students have completed Stage 4 of this series then you can remind them of the similar work that they undertook in the content creation chapter.
2	• Show students the device that they will be using to record video, and show them how to find out the storage capacity.	• If you are using a video recording device that uses an SD card, you should remove this to show the students where it sits in the device. Also point out that the storage capacity is displayed on the SD card so there is a limit to how large their files can be. Point out that each video take will require storage space. • If your device is multi-functional you should show how to find the storage capacity (sometimes this is embossed into the device; other times it can be accessed through a storage menu). Explain that in these devices there are also other items that use up storage capacity, such as the operating system, apps and other files.

Activity	Instruction	Guidance
3	• Demonstrate how to use the device to record a short video. • Demonstrate how to transfer the video to the school computer. • Demonstrate how to delete a video from the device. • Students should complete (📖✏ Page 23) Task A, 'Using the video recorder' to write key notes as you demonstrate the equipment.	• Encourage students to make notes in their Workbooks to remind them how to use the equipment and transfer files.
4	• Talk through the tips for recording a video. • Help students to access their scripts on the computers, or provide them with printed copies if they prefer this method. • Introduce the Build task and guide students to complete (📖✏ Page 24) Task B, 'Video notes' as they rehearse and shoot their takes.	• Tell students that they will have plenty of rehearsal time and time for reshooting their videos. • Mention any extra tips that are specific to your devices or environment. Students can make a note of these in their Workbooks. • Observe students shooting their videos. Remind them of their Workbook notes on how to use the equipment, if necessary. • Students may need help to transfer their final versions to the school network if this is a complex process in your school.
5	• Ask students to complete (📖✏ Page 25) Task C, 'My video'.	• Make sure that all groups have saved their videos to the place you have specified. • Make sure that you can access all videos.
6	• Reflection: In pairs or small groups, ask students to discuss 'Why would it have been easier to record the video in more than one take and then edit it together later?' and then complete (📖✏ Page 24) 'Reflection'.	• Encourage views from all roles. The presenter might have felt that they were only needed for small parts but had to stay alert throughout. The camera person might have felt pressure to know the full script to prepare for each camera movement. The interviewer and interviewee might have felt that time to think between questions would have made sure that all points were presented in the best way. They should also comment on the large file size being a restriction for storage and transfer.

Activity	Instruction	Guidance
	Homework (Optional) (📓 Page 25) • Students are reminded that the videos they recorded may have required several takes. Videos also might need to be cut into smaller segments to remove unwanted pauses or words. Students are asked to watch a local TV news report and note down when they spot key editing features.	
Build on this: Students could revisit their scripts for the final take of their videos to mark points at which the recording could be split. Discuss what they think is the optimal number of smaller segments and whether splitting in this way could introduce more difficulties, such as remembering the order and ensuring continuity.		

Chapter 2: Lesson 6 Broadcast a news report

Student's Book page 36			
Learning Objective/s	• Showcase news reports		
You will need	• Classroom screen to showcase videos	**Key pedagogy**	Classroom talk

Activity	Instruction	Guidance

1	• Showcase the videos one group at a time.
	• Ask 'After each video, discuss: How realistic is the future technology prediction?' (📖 Page 36).
		• Ask students questions such as:
		  • Why did you decide on that particular topic?
		  • How did you work collaboratively on the script?
		  • How easy was it to assign roles and responsibilities within the group?
		• Ask the class how the groups came across as:
		  • knowledgeable on the topic
		  • trustworthy.
		• It might be clear which groups did enough research, believed in the prediction themselves, and picked up on the observations about how to refer to each other as experts.

Activity	Instruction	Guidance
2	• Ask students to complete (Page 26) Task A, 'Reflection'. • Discuss the answers to the Reflection task.	• Encourage all students to join in the reflection discussion. • You could create a class table of pros and cons of online collaboration.
3	• Celebrate the end of project and congratulate all the students on what they have achieved. • Reflection: In pairs or small groups, ask students to share something that they really liked about a news report video made by someone else and then complete (Page 27) 'Reflection'. • Students can then complete the (Page 27) Project skills checklist.	• Encourage all students to share one thing. • Students could comment on the set-up, the camera approach, the content of the interview, switching between speakers and anything else that they think is relevant.
	Homework (Optional) (Page 27) • Students are asked to write down three things that they predict will happen with technology in the future.	

Build on this: Create a class time capsule of their future technology predictions to be opened at a set date in the future.

Chapter 3 – Create with code 1

Project: Design a feedback kiosk using a flowchart and build it in Scratch

Chapter overview

In this chapter, students will discover logic and logic gates, and practise following logic circuits. They will then recall their knowledge of flowcharts and make predictions about their outputs based on inputs. Finally, they will design a feedback kiosk using a flowchart, and build it in Scratch.

Chapter 3: Lesson summary		
Lesson	**Learning Objectives**	**Summary**
1	• Explain how computers consist of logic gates, which operate based on Boolean logic principles • Describe the role of logic gates in circuits, including AND, OR and NOT • Follow and describe logic AND, OR, NOT	Students explore Boolean logic operators using illustrations and an unplugged activity using UNO™ cards. They then learn how Boolean logic is used in circuits through logic gates. They investigate a Scratch project that provides a simulation of the key logic gates AND, OR and NOT.
2	• Predict the outcome of flowcharts that use selection • Follow and use selection statements (IF, THEN, ELSE) in flowcharts • SELECTION in flowcharts is achieved with a decision box	Students are shown that logic operators can be used in program code using a Scratch code example. They are then reminded of the flowchart symbols, through an activity, before making predictions about a flowchart that uses Boolean logic. Next, they are shown a Scratch project that uses Boolean operators to create ingredient combinations for a game. They must predict what will happen and see if they were right by running the project.
3	• Follow and debug algorithms presented as flowcharts	Students are presented with a variety of different flowcharts that each have an error in them. They must spot the error and then rewrite each flowchart with the error fixed.
4	• Design an algorithm that could be expressed as a flowchart • Write appropriate flowcharts for algorithms, limited to sequence and selection	Students design the algorithm and flowchart for their feedback kiosk projects.
5	• Represent a flowchart as program code	Students build their projects in Scratch, based on the design work from last lesson.
6	• Showcase a flowchart and a program	Students showcase their flowcharts and projects to the rest of the class.

End of chapter project	**Design a feedback kiosk using a flowchart and build it in Scratch**
Example ideas that students could design:	Feedback kiosks can sometimes be found in shops or near the exit of an attraction to allow a customer to leave feedback on their experience. Students will design a flowchart for a feedback kiosk and then create a model of one in Scratch.

Why we are learning this	
Useful skills covered in this chapter	• Understanding the logic behind everyday devices such as information and sales kiosks
Links to future careers	• Gain insight into how hardware engineers and software developers design products
Links to other subjects	• Business – gathering feedback on a product or service

Chapter 3 – Create with Code 1

Project: Design a feedback kiosk using a flowchart and build it in Scratch

Name	

	Working towards (1)	Meets expected (2)	Exceeding expected (3)	Score
Flowchart	• A flowchart for the program has been partially created	• A flowchart for the program has been created	• A flowchart for the program has been created and the logic is clear and concise	
Program requirements	• The program meets some of the requirements	• The program meets all of the requirements	• The program meets and exceeds the requirements (e.g. the reaction used animation and sound)	
Test	• Some tests had been carried out	• All tests had been carried out and the reflection task shows an good understanding of their testing	• All tests had been carried out and the reflection task shows an excellent understanding of their testing	

Teacher feedback	
Learner response	

<table>
<tr><th colspan="4">Chapter 3: Lesson 1 Logic gates</th></tr>
<tr><td colspan="4">Student's Book page 39</td></tr>
<tr><td>Learning Objective/s</td><td colspan="3"><ul><li>Explain how computers consist of millions of logic gates, which operate based on Boolean logic principles</li><li>Describe the role of logic gates in circuits, including AND, OR and NOT</li><li>Follow and describe logic AND, OR, NOT</li></ul></td></tr>
<tr><td>New keywords</td><td colspan="3">Boolean logic – Form of algebra that uses the operators AND, OR and NOT
Boolean expression – Expression that either has a TRUE or FALSE outcome
Logic gates – Used in logic circuits in computers. They are based upon the Boolean logic operators AND, OR and NOT</td></tr>
<tr><td>You will need</td><td><ul><li>A set of cards. These could be the printable resource provided or from a card game where cards could be grouped by their characteristics, such as UNO™. Tip: You can find printable card decks online to use in classroom settings.</li><li>Computers with access to Scratch, either online or on the offline/desktop version of Scratch. You will also need the <u>(Website C3L1R1)</u> Scratch Project in a place where your students can access it.</li></ul></td><td>Key pedagogy</td><td>Unplugged</td></tr>
</table>

Activity	Instruction	Guidance

1	• Use the description on page 39 of the Student's Book to explain the meaning of Boolean logic. There are illustrative examples of the Boolean operators AND, OR and NOT to discuss with your students.
	• For the AND example ask: Why is the second image false? They should be able to articulate a response similar to 'both conditions need to be TRUE for the whole expression to be TRUE and the temperature is below 20 degrees so the expression is FALSE'.
	• Ask: 'There are two other possible scenarios for this Boolean expression. What are they?'
	  • Scenario 1: The temperature is less than 20 degrees AND it is not raining. This would give a FALSE outcome.
	  • Scenario 2: The temperature is over 20 degrees AND it is not raining. This would give a FALSE outcome.
	• For the OR example ask: There are two other scenarios for this Boolean expression. What are they?
	  • Scenario 1: The snowman could have a hat and a scarf. This would result in a TRUE outcome.
	  • Scenario 2: The snowman could be wearing a scarf but not a hat. This would result in a TRUE outcome.
		• Use the questioning to guide your discussions with the class. They may not fully grasp the key concepts at this point but should be working towards an understanding by the end of the lesson.
		• You could support students by drawing the other possible scenarios on the board to help them to visualise the conditions and the outcomes of the Boolean expressions.
		• There is another opportunity for students to practise with Boolean expressions in the next activity so don't be too concerned at this point if they don't all understand straight away.

2	• On page 40 of the Student's Book, students are told that their teacher will introduce an activity where they will respond to some Boolean expressions.
	• For this activity you will need cards with numbers and another characteristic such as colours or symbols. The cards must have characteristics that allow them to be grouped and used as part of a Boolean expression. These could be from a game such as UNO™.
	• Printable resource activity: Distribute two cards to each student in the class. Think of some Boolean expressions that can be used on the cards. Students hold their card up if it results in TRUE. For example:
	  • Red AND 5 – If their card is red and has the number 5 then they hold it up.
	  • Black OR red – All students should hold their cards up for this expression.
	  • NOT 7 – Every student with a card that isn't a 7 should hold up their card.
	• UNO™ style example expressions:
	  • Yellow AND 3
	  • Blue OR 1
	  • NOT yellow
		• Keep going with the examples until you are sure that students have grasped the concept of using AND, OR and NOT in a Boolean expression. Their responses should get faster each time, as they become more comfortable with the terms.

Activity	Instruction	Guidance
3	• Ask students to complete (Page 28) Task A, 'Boolean expressions'.	• After giving students a short time to complete this activity, ask them to swap with a partner and use the answer sheet to mark each other's work.
4	• Use pages 40 and 41 of the Student's Book to introduce logic gates to your students. Introduce the symbols and relate them back to the logic expressions that they acted on earlier.	• This is very theoretical knowledge that some students might find tricky to grasp initially. However, the Scratch project 'Logic Gates' should help to develop their understanding in the next task.
5	• Ask students to complete (Pages 28 and 29) Task B, 'Investigate logic gates'. • Demonstrate how to open the Scratch project 'Logic Gates', how to switch to full screen viewing, and how to use the project by pressing the green flag. • Show students that you can click the inputs to switch them ON and OFF. The output of each gate will be a purple dashed line for ON or a grey line for OFF.	• Use the answer sheet provided to work through the answers with your students.
6	• Page 41 of the Student's Book prompts students to think of a Boolean expression that relates to a real-world example. • Ask students to share their real-world Boolean expressions with a partner. They should assess the other person's expressions and ensure that they can only be TRUE or FALSE. • Reflection: Complete (Page 29) 'Reflection'.	• If there is time, give pairs the opportunity to share their expressions with the class. • Examples: "I am NOT wearing a watch" "This classroom has a poster of a tree AND this classroom has a poster of a book" "My chair is blue OR my chair is green"
	Homework (optional) (Page 29) • Students are asked to find examples of Boolean expressions in everyday life.	Examples: • "It is NOT raining" • "I am wearing a sunhat OR I am wearing sunglasses" • "Today is Friday AND the time is after 2pm"
Build on this: Explore the code on the Output sprites to see the Boolean expressions in action. Click 'See Inside' in the Scratch project. Click on each Output sprite to see its code.		

Chapter 3: Lesson 2 Selection in flowcharts

Student's Book page 42

Learning Objective/s	• Predict the outcome of flowcharts that use selection • Follow and use selection statements (IF, THEN, ELSE) in flowcharts • SELECTION in flowcharts is achieved with a decision box		
You will need	• Access to computers that have access to the Scratch project (Website C3L2R1) for students to check their understanding of flowcharts that use selection	**Key pedagogy**	Classroom talk

Activity	Instruction	Guidance
1	• Use the 'What do we already know?' section of page 42 of the Student's Book to remind students of the three operators that they learnt about last lesson. Also use it to remind them of some key facts about flowcharts. This will be recapped during the first activity.	• You may wish to start by asking students if they can remember the three Boolean operators and what they do.
2	• Use the Scratch blocks on page 42 of the Student's Book to show students that Boolean operators can be used in programming as well as logic circuits. The example shown is some Scratch code. State that 'if the time variable is greater than 10 AND the score is greater than 5, then the jump subroutine will be called (run).'	• Some students may have investigated the Scratch code from the 'Logic Gates' Scratch project last lesson. You could point to this as a further example of the Boolean operators being used in programming.
3	• Ask students to complete (Page 30) Task A, 'Label the flowchart symbols' to remind themselves of the symbols that are used for creating a flowchart.	• Use the answers to check students' understanding. You could do this as a peer assessment activity or discuss each answer as a whole class.
4	• Tell students that flowcharts can use Boolean operators too. Show students the flowchart on page 42 of the Student's Book. Ask them to predict the output of the flowchart. The answer is '**Child ticket**'. This is because the **age was 6** and the Boolean expression in the decision symbol evaluated as TRUE.	• You could extend this with another value for score. For example, what if the score was 75? What would the output be? Answer: 'PASS'.

Activity	Instruction	Guidance
5	• Move to page 43 of the Student's Book and introduce the scenario of a student designing a game in Scratch that allows the player to enter ingredient combinations to make some healthy food. • Go through the flowchart with your students and ask 'What will happen if the first ingredient is a tomato and the second ingredient is a potato?' The answer is that the soup bowl costume will be displayed. • Now ask your students to complete (Pages 30 and 31) 'Task B, 'The correct ingredients'. Here they will predict the different outcomes that could occur based on the ingredient inputs.	• Students are developing two skills here. They are recalling how to create flowcharts, use their symbols and follow them through logically. They are also learning how to predict their outcomes based on inputs. This task provides practice for these skills. They will see the flowchart in action as code in the next task. This should help to address any misconceptions that may arise from this part of the lesson. • If some students finish this quickly, they could think of other ingredients that they could input, and then predict the output for this input.
6	• Now that the students have made their predictions in their Workbooks, they can check if they were correct by running the actual Scratch project. • Open the Scratch project on the board. • Demonstrate how to open 'The correct ingredients' project and go through the code blocks on the 'Bowl' sprite. Compare them with the symbols in the flowchart. They should be able to identify similarities. Show students that they need to click the green flag and enter the ingredients from the table into the input box. • They should go through each row in the table and check whether they made the right prediction. They can put a tick or cross next to each answer to mark their work.	• The condition uses OR so the Soup costume will be shown if either the first or second ingredient is correct (or both). • Check that students are using the correct spellings for Tomato and Onion to ensure that the code works as intended.
7	• Reflection: In pairs or small groups, ask students to discuss 'Were any of your predictions incorrect? If so, which ones and what was the actual output?' and then complete (Page 31) 'Reflection'.	• The most challenging prediction is most likely to have been the last one in the table 'Carrot OR Tomato'. This could cause confusion because 'Tomato' for Ingredient 1 is TRUE but it isn't for Ingredient 2. They must look carefully at the whole condition to make an accurate prediction.

Activity	Instruction	Guidance
	Homework (optional) (Page 31) • This asks students to create a new recipe condition for their own soup recipes. They need to remix and modify the Scratch project to use their new conditions.	• You could ask students to share their projects. Check that their code matches the decision they have documented.

Build on this: Extend the Scratch program

Students could modify the Scratch project to include an AND operator instead of an OR operator.

Chapter 3: Lesson 3 Debug flowcharts

Student's Book page 44

Learning Objective/s	• Follow and debug algorithms presented as flowcharts		
You will need	• No special equipment	**Key pedagogy**	Classroom talk

Activity	Instruction	Guidance
1	• Start the lesson with the flowchart discussion activity on page 44 of the Student's Book. A student has designed a flowchart to let them know if they need to wear a raincoat or not. They have implemented the flowchart in Scratch but it is not working as expected. Follow the flowchart through with your class and see if they can spot the error.	• The error is in the decision block and the process blocks that are linked from it. If the output is TRUE, then the raincoat costume should be displayed but it is currently displaying the 'No raincoat' costume. • Students could potentially get confused with why it says 'No coat costume' at the beginning of the flowchart. This is because you want to set the default costume when the green flag is clicked. 'No coat' is the default costume. • 'No coat costume' and 'Coat costume' are outputs.
2	• Now ask students to move to complete (Pages 32 and 33) Task A, 'Debug the flowcharts' where they are given three flowcharts that have errors in them. They should read the description, identify the error with a circle and redraw the flowchart with the error fixed.	• Students could work in pairs for this activity to help provide extra support with debugging.

Activity	Instruction	Guidance
3	• Use the answers provided and the correct versions of the flowcharts to check the work as a class. Solutions to problems when drawn as flowcharts may be different from different people but equally correct.	• There is more than one way to draw a flowchart. Some students might have created slightly different flowcharts to the ones in the answer sheet. Check the logic to see if they are correct. You could make good use of paired work here for students to check the logic in each other's flowcharts.
4	• The flowcharts in the examples only run once. Use the discussion question to ask students what could be added to the flowcharts to make the program constantly check for changes in a condition. The response you are looking for is 'a loop' or 'repetition'. Let students know that they will learn about adding repetition in flowcharts later on in the course. • Reflection: Ask students to complete (📖✏ Page 34) 'Reflection'.	• In this Stage, students are required to understand how selection works in a flowchart. It is important that we don't introduce a misconception here that control devices only act on a piece of code once –they are continually checking for changes so that they can produce the desired output.
	Homework (Optional) (📖✏ Page 34) • Students are given a flowchart for a Scratch project that creates a salad. They need to find the two mistakes and describe how they would fix them.	1. The student has used OR, which means that only one ingredient needs to be correct. OR should be replaced with AND. 2. The student has used 'Tomato' as the second ingredient. 'Tomato' should be replaced with 'Lettuce'. (It's very common to modify a flowchart for a new purpose and forget to make all the required changes for the new scenario.)

Build on this: **The bridge**

Give students an additional scenario for them to create a flowchart for. A bridge is placed over a small river. When a riverboat approaches, the bridge must lift up to let it through. If there are cars waiting to cross then the traffic must be stopped before the bridge is lifted.

Chapter 3: Lesson 4 Design a flowchart

Student's Book page 45

Learning Objective/s	• Design an algorithm that could be expressed as a flowchart • Write appropriate flowcharts for algorithms, limited to sequence and selection

You will need	• Access to the Scratch platform and the sample project (Website C3L4R1).	**Key pedagogy**	Get creative

Activity	Instruction	Guidance
1	• Reintroduce the project brief to the students using the text on page 45 of the Student's Book. Remind students that they will be designing and building a feedback kiosk, and recap what it needs to do. • Go through the worked example with the students. It demonstrates an algorithm for a feedback kiosk. The algorithm is translated into a flowchart, and the flowchart is then converted into program code. • Ask students questions about the similarities between the instructions for each one.	• Worked examples are used in programming to support the development of a mental model of coding patterns. Students will use a very similar structure to the worked example in their own projects.
2	• Demonstrate how to open the 'Rating Kiosk' Scratch project and give students a few minutes to test it out and see how it works. They should enter different ratings to see the costume changes.	• This worked example can be used as a starting point to scaffold the task for students who need it. • Some students will go beyond this and create elaborate animations based on the feedback rating. This is an excellent opportunity for students to be creative with their design decisions.
3	• Students should now be familiar with the task that they will be completing when creating their own feedback kiosks. • Ask them to complete (Page 35) Task A, 'Your feedback kiosk'. This asks them about their design decisions and to implement the code as an algorithm, and then as a flowchart.	• Remind students that there is a worked example on pages 45 and 46 of the Student's Book that they can use as a reference for their own algorithms and flowcharts.
4	• Give students time to reflect on their designs with a partner and ask for feedback. • They should record their feedback in (Page 35) 'Reflection'.	• You could implement a process of 'two stars and a wish' by asking students to state two things that they like about their idea and one thing that they wish it could have.

Activity	Instruction	Guidance
	Homework (Optional) (📖✏ Page 35) • This asks students to find an example of a real feedback kiosk or a feedback option for a computer app or website. They need to describe how this example gathers feedback and how effective they think this is.	• Mention any feedback kiosks that students may be familiar with. This could be in the school, the local community or a place that students have visited on a school trip. • Alternatively, many apps and websites have an option to provide feedback. • Feedback can be captured through stars, numbers, emoji faces or text statements.

Build on this: **Action feedback**

Students could action their partner's feedback.

Chapter 3: Lesson 5 Build a program from a flowchart

Student's Book page 47

Learning Objective/s	• Represent a flowchart as program code		
You will need	• The designs from the last lesson that are in the students' Workbooks. • Access to the Scratch platform and the ability to save projects.	**Key pedagogy**	Get creative

Activity	Instruction	Guidance
1	• Remind students that last lesson they designed the flowcharts for their feedback kiosks. This lesson they will use their designs to build their projects in Scratch.	• Remind students that there is a worked example in the Student's Book if they need additional support.
2	• Instruct students to open Scratch and begin creating their projects. You should remind them of how to save their projects in your chosen method.	• Use the information at the beginning of the teacher guide to support you with saving work. • Sample code is available for students in the Student's Book on page 46 if they need additional support with this.

Activity	Instruction	Guidance
3	• After they have completed their projects, they should test them by completing (📖 Page 36) Task A, 'Test your program'.	• This is a simple test plan for each number that could be entered by the user. Students should record what the sprite does for each number entered and see if that was the reaction they expected.
4	• Give the students some time to reflect on their testing process and whether it showed any errors. They should discuss this with a partner or the class. • They should then record their reflections in their Workbooks (📖 Page 36) 'Reflection'.	• Refer back to the worked example code if students have issues with their code not working as expected. • Ask students questions to find out the importance of testing the program close to the boundaries of any conditions to make sure that the parameters are correct, e.g. if < 5 then 4, 5 and 6 would all be boundary data to use to check it is correct.
	Homework (Optional) (📖 Page 36) • Students are asked to show their projects to a friend or family member and make notes for improvements based on the feedback they get.	• Students may find mistakes that they didn't find in the lesson. They may get feedback on how to make their project easier for a user to understand.

Build on this: Fix errors in their code

If their test plans showed that their code had an error then they should work on fixing those errors if there is time to do so.

Chapter 3: Lesson 6 Showcase a flowchart and a program

Student's Book page 48

Learning Objective/s	• Showcase a flowchart and a program		
You will need	• The flowcharts from the design work in Lesson 4. • The Scratch projects based on the flowcharts from Lesson 5.	**Key pedagogy**	Showcase

Activity	Instruction	Guidance
1	• Demonstrate how to open the Scratch projects from the last lesson.	• See the teacher guidance at the beginning of this book for support with this.

Activity	Instruction	Guidance
2	• Remind students what a showcase is (📖 Page 48). and give them time to practise what they want to say. • They should showcase their flowcharts as well as their completed projects.	• You could get students to share their projects so you can show them all on a computer connected to a projector.
3	• Students showcase their projects.	• Use the showcase section at the beginning of the teacher guide to decide how best to run the session.
4	• Ask your students to reflect on their projects by completing (📖 Page 37) Task A, 'Reflection'.	• Make it clear that it is common to make mistakes when you are planning or writing code. Debugging is a normal part of programming.
5	• Give students some time to share their reflections with their peers (pairs or small groups).	• Encourage students to share their mistakes as well as their successes. You often learn more from understanding what can go wrong when creating a program.
6	• Reflection: In pairs or small groups, ask students to 'Share your favourite project made by someone in your class', and then complete (📖 Page 37) 'Reflection'. • Students can then complete the (📖 Page 37) Project skills checklist.	
	Homework (Optional) (📖 Page 37) • Students are asked to identify any additional improvements that result from presenting their showcase or observing other students.	• It's common to think of something you would like to change when you show a project to others.

Build on this: **Action feedback**

Students could improve their feedback kiosks based on the comments from the audience of their showcase.

Chapter 4 – How computers work

Project: Create a collaborative pixel art installation that represents your local community

Chapter overview

In this chapter, students will look at the technology around them with a focus on automation in the health and advertising industries. They will then learn about applications and systems software and the differences between them. Next, they will discover how media is represented in binary before collaborating on a pixel art installation.

Chapter 4: Lesson summary		
Lesson	**Learning Objectives**	**Summary**
1	• Compare application software with system software • Explain how automation is used in the health and advertising industries	Students are introduced to the two categories of software: application and system software. They are given examples and complete a sorting activity to assess their understanding. Next, students discover how software can be used to support automation in the advertising and healthcare industries. Then they have a class debate on whether automation is good or bad.
2	• Describe how numbers, characters and sound are represented in binary	Students are presented with background knowledge on how binary is used to represent numbers, characters and sound. There are practical applications of this through Workbook activities.
3	• Describe how images are represented in binary • Research images that represent the local community	Students explore how images are represented in binary using examples of 1-bit and 2-bit colour depth. They are then given time to discuss and explore images that represent their local community.
4	• Design a piece of pixel art for an installation	Students are given time to develop three designs for their pixel art.
5	• Create a piece of pixel art for an installation	Students create their final pixel art designs.
6	• Showcase a piece of collaborative artwork	Students take their pieces of pixel art and join them together with the other pieces of art that their classmates have created to form a large art installation. They then showcase their individual pieces, explaining how they think it represents their local community.

End of chapter project	**Create a collaborative pixel art installation that represents your local community**
Example ideas that students could design:	A community is personal to individuals. It could be art that represents your class, your school or your local area. There might be a famous landmark nearby or school symbol or motto that students want to represent. Encourage students to take some time to research and think about the types of things your community represents. 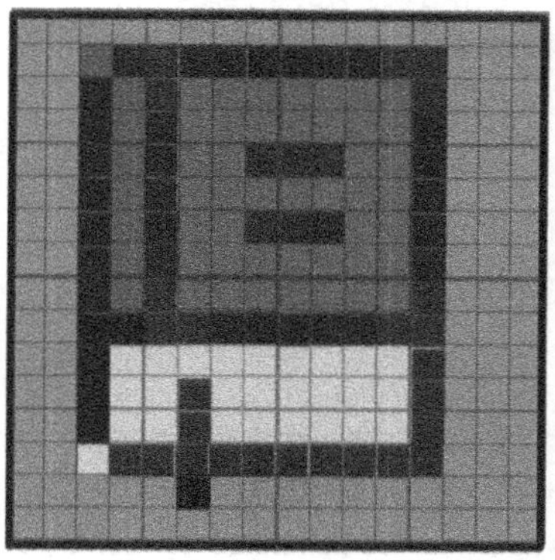  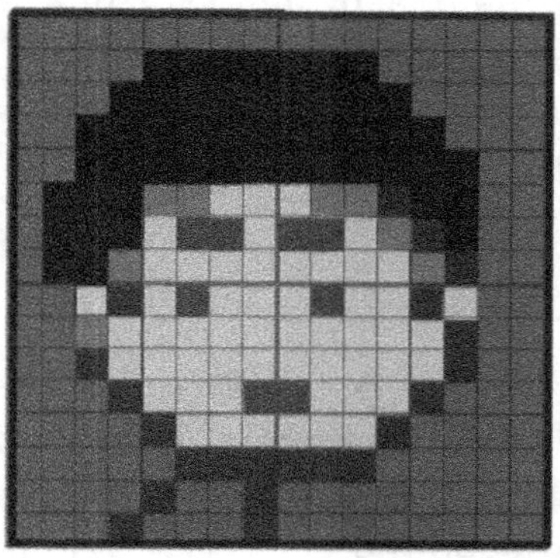

Why we are learning this	
Useful skills covered in this chapter	• How automation is used in advertising and healthcare, how data is represented in binary, how to create a pixel art image
Links to future careers	• Digital artist, software engineering
Links to other subjects	• Art, Science (sound waves)

Chapter 4 – How computers work

Project: Create a collaborative pixel art installation that represents your local community

Name	

	Working towards (1)	Meets expected (2)	Exceeding expected (3)	Score
Design work	• Minimal design work has been carried out	• There are three draft designs for the pixel art	• There are three draft designs for the pixel art and the design decisions are well described in the reflection	
Colour palette	• The colour palette has not been used for the final design	• A colour palette has been used for the final design	• A colour palette has been used for the final design and the choice of colours was meaningful (e.g. I chose green and yellow because these are the school colours)	
Represent the local community	• There is a minimal connection between the image and the local community. The learner may not be able to describe the link fully.	• The image represents the local community and the learner provides a good description of how it links.	• The image represents the local community and the learner provides an excellent description of how it links.	

Teacher feedback	
Learner response	

Chapter 4:	Lesson 1 Software and automation		
Student's Book page 51			
Learning Objective/s	• Compare application software with system software • Explain how automation is used in the health and advertising industries		
New keywords	**Application software** – Software used for a specific task, such as a word processor **System software** – Software that controls how the computer operates **Operating system** – Software that manages the hardware, software and resources of a computer **Utility software** – Software that provides additional support for a system, such as antivirus or disk cleanup management **Automation** – Carrying out tasks with little to no human interaction		
You will need	• Not applicable	**Key pedagogy**	Real-world contexts

Activity	Instruction	Guidance
1	• Introduce the project the students are working towards (Page 50).	• If students have learnt about mosaic artwork in the art curriculum, or perhaps when studying the Romans, you could emphasise a link between that and pixel art.
2	• Introduce the term 'application software' to students, and provide examples using the list (Page 51). • Ask 'Can you think of any more examples of application software?'	• Think about the applications that are installed on your school computers. Which ones haven't been listed in the examples?
3	• Introduce the term 'system software' to the students (Page 52). There are fewer examples of system software because it is very specific to the type of device that you are using. In general, computing devices will have an operating system as the prime example of system software. Some will also have utility software, such as anti-virus, restore and backup, and disk cleanup.	• System software interacts directly with the hardware components. Application software sits on the layer above that; it will interact with the system software, which will then interact with the hardware on its behalf.
4	• Give students a few moments to complete the table on (Page 38) Task A, 'System or application software'. They should tick the relevant column to state whether the example is system or application software.	• You could extend this task by asking students to note down further examples that they have on the school computers.

Activity	Instruction	Guidance

5	• Explain what automation is and give examples of its use in healthcare (📖 Pages 52 and 53).
	• Ask 'Do you think that it is a good idea to leave automation software to make decisions about the urgency of a patient's illness?'
		• There are two sides to this discussion question. This discussion will help form the debate points in a later activity.
		• Positives:
		  • It saves time because scheduling can be done without the need of an administrator.
		  • The appointment scheduling might be fairer because the system would not allow patients to jump the queue. There is no bias in an automated system.
		• Negatives:
		  • There may be errors in the tool that stop it from assessing the urgency of the appointment.
		  • A patient could be put at risk if they are unable to explain their illness through an automated tool.

6	• Continue to discuss automation and introduce examples of automation in the advertising industry (📖 Page 54). • Ask 'Have you ever been on a webpage and seen an advert that is linked to what you have recently searched for? Do you think that this is good or bad?'	• Students may have already seen examples of targeted advertising when they have been searching online. Remind them that cookies are used to track what you are doing online and target advertising based on your search history.
7	• Ask students to answer the questions in the (📝 Pages 38 and 39) Task B, 'Is automation good or bad?' They should list two reasons why automation is good and two reasons why automation can have negative effects. • Let students know that these points will be used in a debate that they will be having in the next activity.	• They can use the information in the Student's Book to support this. They can also add any points from the discussion activities that they think are relevant.

Activity	Instruction	Guidance
8	• Lead a class debate on the benefits and drawbacks of automation. You could run this session as a whole class or in small groups. Randomly allocate members of the class to each side. They must use the reasons in their Workbooks to guide their point of discussions. • After the debate, ask your whole class to vote using a show of hands on whether they think automation is good or bad.	• Some students will feel more comfortable debating in small groups or pairs. You could give them some time to practise or write down some key points prior to asking them to address the whole class.
9	• Reflection: In pairs or small groups, ask students to discuss 'Which side did you agree with the most? Can automation only be good or bad? Could it be both?' and then complete (Page 39) 'Reflection'.	• Students can think about the side that they agreed with most, but they should also be working towards the conclusion that technology can be used positively and negatively. It isn't typically just good or bad.
	Homework (Optional) (Page 39) • This asks students to consider examples of advertising in the world around them.	• Suggest typical places where advertising might be shown to students in your local community. For example, it could be through streaming services, newspapers, billboards or television.
Build on this: Research the advertising or healthcare industries further. If you have access to devices with the internet, then students could spend some time researching more about the advertising and healthcare industries, and how they use automation tools. You can link back to Chapter 2, where they learnt how to search for trustworthy information online.		

Chapter 4: Lesson 2 Binary representation

Student's Book page 55			
Learning Objective/s	• Describe how numbers, characters and sound are represented in binary		
New keywords	**Decimal** – Base-10 number system that uses 10 digits, from 0 to 9 **Denary** – Alternative for 'decimal' **Analogue sound** – Sounds that are all around us in the non-digital world **Characters** – Numbers, letters and punctuation, for example, A, " or 9		
You will need	• The (Website C4L2R1) Scratch project	**Key pedagogy**	Real-world contexts

Activity	Instruction	Guidance
1	• Remind students that they should already know that computers represent data using 1s and 0s. • Introduce them to how binary is used to represent numbers (📖 Page 55).	• This is quite a tricky concept to grasp. Don't expect all students to take this in at the first explanation. The Scratch project in the next activity will provide another way to look at the concept, and there is a further activity in their Workbooks to consolidate it.
2	• Open the 'Count in binary' Scratch project on your teacher machine and view it in FullScreen mode. Press the green flag to start the project. • Click on the 'Count up' button to slowly step through how the decimal numbers from 1 to 10 are represented using binary.	• Keep slowly counting up with your students to help them to develop their understanding of how the numbers 0 to 10 are represented in binary. • If there is time in the lesson, then you may wish students to access the project themselves and go through it slowly. • Another activity could be to ask students to think how they might write the decimal number '11' in binary.
3	• Instruct students to complete the table on (📖✏ Page 40) Task A, 'Count in binary'. They are asked to write the numbers 0 to 10 in binary.	• If students require additional support here, then you could provide access to the Scratch project. • Once students have completed the task, use the answer sheet or the Scratch project to help them to peer mark their work.
4	• Introduce the students to how binary is used to represent the characters on a keyboard (📖 Page 56).	• The examples use a standard English or American keyboard. The same principle applies to keyboards that use non-Latin characters, but these will have different binary codes to represent them.
5	• Now give a brief explanation of how sound is represented in binary (📖 Page 57). • You may wish to link the topic of sound waves with students' science lessons.	• Note that students will cover more detail on how sound is represented in Stage 9.

Activity	Instruction	Guidance
6	• Ask students to complete the activities on (Page 41) Task B, 'Represent characters in and (Pages 41 and 42) Task C, 'Represent sound in binary'. • First, they are provided with a table showing letters from the English alphabet and the binary codes used to represent them. They are given three words to write in binary code using the table. • The second task asks them to take samples from an analogue sound wave and digitise them using a binary code. They can mark the image with an X by following the time at the bottom up to where it intersects with the binary value on the left. The first three have been marked to show them what to do. • Once students have finished, you could go through the answers to the sound task and discuss why some students might have slightly different answers due to how they interpreted the wave.	• For the characters, advise students to choose shorter names if you think this type of activity might take them a while. For example, suitable words might be FISH, RED, BIKE. • For Task C, note that some students may have placed their crosses in a row higher or lower than shown in the answer image. For example, at 'Time 7' the cross could be on 0110 and not 0101. This doesn't mean that they are incorrect, they are just a human doing a computer's job. The computer would be more accurate. This is a good opportunity to point this out to students and could be a nice discussion point.
7	• To lead into the next lesson, have a quick discussion of how a binary number could be used to represent a single colour. They will learn about this in the next lesson. • They should record their thoughts in (Page 42) 'Reflection'.	• The focus here is on a single colour, not the entire image. Perhaps give them four colours and ask them to assign a unique binary code to each of those colours.
	Homework (Optional) (Page 42) • This asks students to write their names in binary.	

Build on this: **Create secret messages**

Students could use the character table in their Workbooks to write a secret message in binary for another person in the class to decode.

Chapter 4: Lesson 3 Represent images in binary

Student's Book page 58

Learning Objective/s	• Describe how images are represented in binary • Research images that represent the local community

New keywords	**Colour depth** – Number of bits (1s and 0s) that are available for each pixel. The higher the number of bits, the more colours that are available **Pixel** – Block of colour that forms part of an image. A binary code is used to represent each colour		
You will need	• A selection of coloured pencils to distribute for the workbook activity (optional) • A collection of ideas for images that represent your local community. You could restrict this to things that represent your school, or you could widen it to your local area. This would work really well if you have a specific landmark nearby, or your area is famous for something that students could make an image of.	**Key pedagogy**	Get creative

Activity	Instruction	Guidance
1	• Remind students that computers represent data using 1s and 0s, and recall the last lesson where they discovered how numbers, characters and sound were represented.	• You could ask a specific question, such as 'How are numbers represented in binary?' They should answer by saying that they use the binary digits 0 and 1 only. They may expand on this by stating that binary uses a base-2 number system.
2	• Introduce the concept of pixels in an image. The examples and explanations will support this (Page 58). Use the example graphic to demonstrate that if you zoom right into a digital image, you can see the individual pixels. • Explain that each pixel is a block of colour that has a unique binary code. Use the example blue and white graphic to demonstrate how a 1 and a 0 can be used to represent each colour to make up the image (Page 58).	• Additional knowledge – a 'bit map' is created for the image to map out where each bit will appear in the grid. Metadata is used at the beginning of the binary encoding to state the resolution and bit depth of the image so that the data can be processed and displayed as an image on the screen. Students do not need to go into this level of detail at this stage but you may wish to have this information at hand in case students ask for it.

Activity	Instruction	Guidance
3	• Ask students to complete (📓✎ Page 43) Task A, 'Draw the 1-bit colour depth images', where they will look at the binary codes and colour in the squares to match the binary digits.	• The answers have been provided. These can be used to check students' understanding. They may potentially mix up the numbers and use a 0 for the colour and a 1 for white. If students are finding this activity challenging, then you could draw a grid on your board and complete the first example with the class (or a small group).
4	• Now introduce students to the concept that if they have more bits, or a greater bit depth, then they can represent more colours (📖 Page 59). The example graphic shows a 2-bit colour depth image that uses four colours (because you can make four different combinations of 1s and 0s with 2 bits).	• For simplicity, any colour has been assigned to a bit pattern. As students discover more about data representation, later on in Computing, they will learn about hex codes that are common in web design and other forms of design, and use specific codes for specific colours.
5	• Ask students to complete (📓✎ Page 44) Task B, 'Colour combinations'. They must create as many combinations of 1s and 0s as they can, first for 3 bits and then for 4 bits. Do explain that the best way to do this is to count up from 0 in binary. • The answers are provided. You may wish to ask students to peer mark to check their answers.	• There is a tip in the workbook for students to look back at their 'Count in binary' work from the last lesson. If they count from 0 to 8 in binary using 3 bits, they will reveal all combinations that use 3 bits. The same applies to the next question with 4 bits. If they count up to 16 using 4 bits, they will reveal all of the bit patterns in order. • If students struggle with this, you could work out the 3-bit question together as a class. They could also work in pairs to check for duplicates.
6	• Now remind students of the project brief for this chapter. They will be each contributing one piece of pixel art to a large art project that will be created by the whole class. • Discuss your local community and the types of graphics that could be used. If you are basing it on your class, perhaps they could do self-portraits or an image that represents a favourite hobby.	• You may want to draw some specific examples from your local area as inspiration for your class.

Activity	Instruction	Guidance
7	• As part of your discussion work, students could record sketches and ideas on (📖✏ Page 45) Task C, 'Images that represent my community'. They could work in pairs for this or discuss their notes with a partner, or the class, to end the lesson.	• Next lesson, students will have the whole time to draw three different sketches for their chosen image or images. The sketches that they do now should be rough ideas.
8	• Reflection: In pairs or small groups, ask students to discuss 'What image do you think you might use in your project?' and then complete (📖✏ Page 45) 'Reflection'.	
	Homework (Optional) (📖✏ Page 45) • This asks students to complete the 'Spot the pattern' activity.	

Build on this: **Review and extend**

If there is time remaining, students could review each other's basic sketches and ideas, and add to them after a short discussion.

Chapter 4: Lesson 4 Design your pixel art

Student's Book page 61

Learning Objective/s	• Design a piece of pixel art for an installation		
You will need	• A collection of coloured pencils or felt-tip pens. There should be a minimum of 8 colours, ideally including black.	**Key pedagogy**	Get creative

Activity	Instruction	Guidance
1	• This lesson is focused on students designing three versions of their community pixel art. Remind them of the key points of the project (📖 Page 61). • Next, show them the three colour palettes that are used as examples. These are all slightly different but show good choices that would make it easier to draw an image.	• The colour palettes are there to show students that they shouldn't go for the obvious 'colours of the rainbow' choices. Having a rainbow colour palette can make it quite difficult to draw a pleasing image. It is better to have more subtle colour palettes that use shades of similar colours and some contrasting colours. • (Website C4L4R1) is a great website for generating different colour palettes

Activity	Instruction	Guidance
2	• Give students the majority of the lesson to develop their three design ideas using (Pages 46 and 47) Task A, 'Design three versions of your pixel art'. • They may need longer than this and could use a bit of time in the next lesson as well, if needed.	• It is better to design the colour palette first, and then use that colour palette to draw the image before adding in the binary codes. • They may find that they adjust their colour palette as they go. This is normal and should be encouraged.

3	• Set some time aside at the end of the lesson to reflect on their designs. Ask:
	  • Do you have a favourite design from the three that you drew in this lesson?
	  • Were you happy with your colour palette?
	  • Did you have to change your colour palette as you were drawing the image?
	• Ask students to record their reflections in (Page 47) 'Reflection'.
		• Remind students that design is a process that constantly changes and improves as they work on their images. There shouldn't be a 'perfect' result the first time.

	Homework (Optional) (Page 47) • Students are asked to consider the colour palettes used in some company logos.	
	Build on this: **Review** Students could share their designs with a partner and discuss which one is the best. They could also give each other feedback on their designs so that they can make adjustments next lesson.	

Chapter 4: Lesson 5 Create your pixel art

Student's Book page 62

Learning Objective/s	• Create a piece of pixel art for an installation		
You will need	• Coloured pencils, preferably the same options as last lesson • Space has been provided in the Workbook for students to draw their final designs. These could then be scanned/colour photocopied or cut out to form part of the larger artwork. If you don't want to cut sections from the Workbook then you should provide squared paper or a larger grid of 16 × 16 squares for students to use. for their art piece.	**Key pedagogy**	Get creative

Activity	Instruction	Guidance
1	• Remind students of the three designs that they made last lesson. They should have already chosen their final design, but they may need a short amount of time to work on their draft design work.	• You may wish to provide additional pieces of squared paper if students want to do some further drafting before choosing their final design.
2	• Give students time to create their final designs. There are two options here. The first is to draw their final designs in their Workbooks, using (Page 48) Task A, 'Draw your final design', and then scan or colour photocopy them for the final piece. The second option is to provide the students with squared paper on which to draw their final designs so that these can easily be cut out and linked together with the other designs in the class.	• There are also online pixel art tools that students might wish to use instead. They could recreate their final design using an online tool. • (Website C4L5R1) allows you to set the canvas size to 16 × 16 and choose your colour palette. • Make sure that you provide the same coloured pencil options as last lesson to ensure continuity of colours.
3	• Ask students to reflect on their final designs. Ask 'How does your artwork represent your school community?' This will help them form their script for their showcase in the next lesson. • Ask the students to record their reflections in (Page 48) 'Reflection'.	
	Homework (Optional) (Page 48) • Students are asked to look for website Favicons and document their findings.	
	Build on this: **Review** Students could share their final designs with a peer and ask for feedback.	

Chapter 4:	Lesson 6 Showcase your artwork		
Student's Book page 63			
Learning Objective/s	• Showcase a piece of collaborative artwork		
You will need	• The final artwork designs that students created last lesson • Sticky tape if you wish to permanently join the art pieces together. Another option is to pin the pieces to a display board.	**Key pedagogy**	Showcase

Activity	Instruction	Guidance
1	• Students should now have their completed piece of artwork. Ask them to cut out their designs so that it just leaves the square that they will add to the class piece.	• You may wish to do this in advance of the lesson to avoid any mishaps.
2	• You could split the class into groups of four or six and ask them to decide on the best layout for each of their sections of the larger piece of artwork. Once they have decided how to lay them out, they should flip the designs over and use sticky tape or glue to carefully create a larger square of their designs. • Once each group has created their larger square, you can take each square and join them together to form the larger art piece.	• Option: You may wish to pin or staple the designs to a display board instead of asking the students to do this themselves.
3	• Note: This step could be done before joining the artwork together to save time if needed. • Give students time to practise what they want to say about their own art piece. • Ask students to clearly showcase their part of the art installation. If they have practised, then they should be able to do this as a whole class. They should clearly show which part of the art installation is theirs, what colour palette they chose and why, and explain how their image reflects their local community.	• A nice addition here would be to film their showcase for a school assembly or for social media, if you have the right permissions from parents and carers to film and display publicly.
4	• Give students time to appreciate the collaborative piece of artwork that they have all created together, and praise them for their achievements.	• Give them a few moments to sit and look at what they have achieved.
5	• Ask students to complete (Page 49) Task A, 'Reflection'. • Share their reflections with a partner or small groups. • Reflection: In pairs or small groups, ask students to discuss 'What was it like to see your piece of artwork as part of a collaborative installation? How well did your class collaboration represent your community?' then complete (Page 49) 'Reflection'. • Students can then complete the (Page 49) Project skills checklist.	

Activity	Instruction	Guidance
	Homework (Optional) (📒✏ Page 49) • Students are asked to research some famous pixel art images.	• Think about the types of images that your students may have seen in your community. For example, they may have seen old video games where the characters were made using simple pixel art, such as Super Mario.

Build on this: **Showcase to a larger audience**

If there is curriculum time available it would provide a great opportunity for students to showcase their artwork to a wider audience. This could be in an assembly, or you could invite parents and carers to come and view the work at the end of the school day.

Chapter 5 – Create with code 2

Project: Make a recommendation system

Chapter overview

In this chapter, students will be introduced to the text-based programming language Python through the Scratch-like EduBlocks tools. They will learn how to create text-based programs that use input and output, variables, operators, conditional statements and functions. They will apply their skills to create a text-based recommendation system that asks the user questions and then makes a recommendation based on their answers.

Chapter 5: Lesson summary		
Lesson	**Learning Objectives**	**Summary**
1	• Develop text-based programs that use input and output • Use variables in text-based programs	Students will recall computer programming concepts and write their first programs in Python using the block-based EduBlocks tool. They will create a silly sentences program using variables and input and output.
2	• Identify and define data types in text-based programs, including Integer, Real and String • Develop text-based programs using data types, including Integer, Real and String • Develop text-based programs that use different arithmetic operators (+, -, *, /) • Use appropriate comparison operators (<, >, <=, >=, ==, !=) in algorithms	Students will identify data types in Python and use their knowledge to create programs that work correctly with inputs of different data types. Students will use arithmetic and comparison operators in Python programs. They will use 'if' statements to make decisions based on the result of comparison operators.
3	• Add effective comments to a text-based program • Recognise patterns in algorithms and edit to use subroutines • Follow and debug algorithms that use subroutines	Students will identify patterns in their code that would benefit from the use of a Python function. They will define a Python function and add a parameter, and call the function from multiple places in their code. They will debug their program so that it works correctly. Students will also add effective comments to their Python program.
4	• Create a text-based program that uses input, output, comparison operators and functions	Students will be introduced to the project brief of creating a recommendation system by making decisions based on a user's answers to questions. They will plan their project and write or modify code to get the input and make a decision.

5	• Create a text-based program that uses input, output, comparison operators and functions	Students will add or modify a function that will provide the shared code for outputting a formatted recommendation. They will use input parameters to provide the data that should be different for different function calls.
6	• Read and explain code written by another programmer	Students will share their projects. They will try projects created by others and give helpful feedback. They will also read another student's code and summarise it to demonstrate their comprehension.

End of chapter project	**Make a recommendation system**
Example ideas that students could design:	Make a recommendation system that asks the user questions and then makes a recommendation based on their answers. The inputs should include at least one text (str) and at least one numeric data type (int or float). Example projects that students could create include: • a recommendation system that recommends a book to read based on the user's age and preferences • an activity recommendation system for your local area that makes a recommendation based on the temperature and the user's interests • a soft drink recommendation system that recommends a drink based on a user's preferences (for example, banana vs strawberry) and budget (for example, small vs large) • an exercise activity depending on level of physical activity wanted (number on a scale) and indoor or outdoor preference • a game upgrade recommendation system based on preferences and number of coins or points available.

Why we are learning this	
Useful skills covered in this chapter	• Understanding how recommendation apps might be created in the real world
Links to future careers	• This chapter links to software engineering and user experience design
Links to other subjects	• There are links to data types and algebra in Mathematics

Chapter 5 – Create with code 2

Project: Make a Recommendation System

Name	

	Working towards (1)	Meets expected (2)	Exceeding expected (3)	Score
Designed a recommendation system	• Needed significant assistance to identify conditions and variables	• Designed a recommendation system including identifying conditions and variables	• Introduced additional conditions/variable s to create a recommendation system with additional options	
Programmed a recommendation system	• Partially programmed a recommendation system that meets some of the project criteria	• Programmed a recommendation system that meets at least four of the project criteria	• Programmed a recommendation systems that meets all of the project criteria and/or introduces additional features	
Identify and debug errors	• Identified errors but needed assistance to debug	• Identified errors with some success at debugging	• Identified errors and explained their approach to debugging	

Teacher feedback	
Learner response	

Chapter 5:	Lesson 1 Python: input and output		
Student's Book page 66			
Learning Objective/s	• Develop text-based programs that use input and output • Use variables in text-based programs		
New keywords	**Text-based programming language** – Allows you to type text instructions that a computer can run **Python** – Text-based programming language that is popular in industry **Subroutine** – Named section of code that can be called and performs a specific task **Function (Python)** – Type of subroutine that returns a value. Python has built-in functions that are premade for you to use. You can also write your own functions in Python **Call (a subroutine)** – To tell a program to execute the code within a subroutine **Argument** – Input to a subroutine call **Statement** – Complete instruction that the computer can run		
You will need	• Student access to the EduBlocks website (Website C5L1R1) • EduBlocks accounts for students (optional, but recommended) • EduBlocks 123 Coding Cards (printed out or link shared with students) (Website C5L1R2)	**Key pedagogy**	Building on prior knowledge

Activity	Instruction	Guidance
1	• Introduce the new chapter. Explain that you will be learning a new programming language and then using it to make a project. • Provide students with the EduBlocks Coding Cards that map from Scratch to EduBlocks to Python. These can be used throughout this chapter. • Cover the material on text-based programming languages, Python and EduBlocks (Pages 66 and 67). • Ask 'Can you spot the similarities between the code on the left and the code on the right? What are they?'	• Make connections to previous computer programming in Scratch. • The answer to the discussion question is that they both display the text '# Start code here' and 'print("Hello World"); however, one is in block format and the other is in text format.

Activity	Instruction	Guidance
2	• Move on to the 'Use **`print`** to output text' section (Pages 67 and 68). Remind them that in Scratch you can create subroutines using the My Blocks feature. • State that in Python, subroutines are called functions. • Use the diagram in the book to clarify the new terminology 'function name/call', 'statement' and 'argument'. • Use the discussion question to ask students to predict what text will be displayed on the screen when the code is run. Ask them why they think that.	• They recapped subroutines in Chapter 3. If they followed this course in earlier stages, they will have used subroutines in Scratch in Stage 6. • Discussion question: don't reveal the answer just yet as they will see the answer when they run the code themselves.
3	• Demonstrate using EduBlocks to create the Hello World example (Page 68). • Point out the text Python code that appears on the right when you drag in the blocks. • Run the example. • Ask 'Was your prediction correct? How could you adapt the code to display a different message?' • Point out that the EduBlocks editor now has a new button for 'Output' as well as 'Code'. They can switch between the two to see the output of the program or the Python code. • Give them some time to adapt the code to display a different message. • When finished, they should move on to (Page 50) Task A, 'Print statements'.	• To adapt the code to display a different message, students change the words inside the " " quotes. • Use the vocabulary of 'print statement' as this will help students to find the print block in EduBlocks. • Note that blocks are colour-coded to help you to find them. • Mention that EduBlocks helps us to create correct Python code. The Python code must be correct to ensure it runs as we want it to. By correct we mean that code must follow the syntax (rules) of the language • Point out the quotes and the lowercase 'p' in 'print'.

4	• Demonstrate using EduBlocks to create the 'Ask for input' example (Page 69).
	• Draw attention to:
	  ◦ creating a variable
	  ◦ dragging the input block into the assignment.
	• Ask 'What do you think will happen when the code is run? Why do you think that?'
	• Demonstrate running the code to show what happens. You will be able to type your name but once your press enter, it will look like nothing has happened.
	• See if their prediction was correct.
	• Ask the students to try it themselves.
		• Relate the code to ask/answer in Scratch.

Activity	Instruction	Guidance
5	• Move on to the 'Use the print function to display a greeting' section (Page 69). • Talk through the new line of code that has been added to the program. • Ask 'What do you think will happen when this code is run? Why do you think that?' • Demonstrate adding the second line of code to your students. Note that a purple block now sits inside the print statement. Purple blocks are 'text' blocks. • Run the code and demonstrate typing your name. A greeting will now appear once you press the enter key. Was their prediction correct? • Give the students some time to try building the code themselves.	• You might want to pause here and also point out the text version of the Python code that has been created from the blocks. This will help develop some familiarity with the text to support them in the next stage.
6	• (Page 70) Run the silly sentences program (Website C5L1R3). • Ask students to provide answers to the questions. • Repeat to create silly sentences. • Walk through the code. • Ask students to create their own version using (Pages 50 and 51) Task B, 'Silly sentences'.	• Make sure that students' suggestions are appropriate. • You may get sentences that are not grammatically correct – explain that Python just does exactly what the program tells it. They will learn more coding skills to write better programs throughout the course. • Students may want to save their projects and create new ones. Show them how to do this. • Observe students and guide them to choose an appropriate number of blanks (variables) so that they will have time to write the code.
7	• As students complete the task, put them into pairs and get them to try each other's programs. • Encourage students to be supportive of each other and help to fix any bugs they find.	• This activity is an opportunity for students to gain confidence in using EduBlocks and quickly finding and connecting blocks. • Continue to observe and assist. • Students who complete the activity can make their example more complex, or create another example. • Keep a class list of error messages and solutions in a shared document or on a whiteboard.

Activity	Instruction	Guidance

8	• Ask 'In what ways is Python similar to Scratch? In what ways is it different?'
	• They should record their reflections in (Page 52) 'Reflection'.
		• Answers could include:
		  • EduBlocks makes it more similar (make sure students realise that Python is text-based and EduBlocks helps them to enter Python).
		  • 'print' is like 'say'
		  • '=' is like 'set'
		  • 'input' is like 'ask'
		  • In Python you have to create a variable for the result of input and assign it, whereas Scratch creates the 'answer' variable for you.
		  • Python is used in industry.

	Homework (Optional) (Page 52) • Students are asked to state the order for the blocks of code to create a program that asks a user for their favourite colour and provides some feedback.	• Go through the answers with the students to check understanding.

Build on this: Write your own story

Ask students to write their own story using the EduBlocks and Python skills they have learnt. They could use their sentence as the first line of the story and then ask the user for more input.

Chapter 5: Lesson 2 Python: operators, conditions and data types

Student's Book page 71

Learning Objective/s	• Identify and define data types in text-based programs, including Integer, Real and String • Develop text-based programs using data types, including Integer, Real and String • Develop text-based programs that use different arithmetic operators (+, -, *, /) • Use appropriate comparison operators (<, >, <=, >=, ==, !=) in algorithms		
New keywords	**Data type** – Classification such as string or integer that tells the computer how to work with a value **Integer** – Whole number		
You will need	• Student access to the EduBlocks website (Website C5L2R1) • EduBlocks accounts for students (optional, but recommended)	**Key pedagogy**	Investigate

Activity	Instruction	Guidance
1	• Remind students of arithmetic operations and demonstrate entering the example code in EduBlocks (Page 71). • Define data types (Page 71) and introduce (Page 53) Task A, 'Why do we need data types?'.	• Remind students that they created a variable last lesson, and complete a reminder demonstration if you think this is needed. They will need to create an 'a' and 'b' variable to complete the Workbook task. • Relate operators to maths and remind students that they have used similar blocks in Scratch. • Point out that the operator can easily be changed from the drop down in EduBlocks.
2	• Ask 'Can you explain the result you got?' The answer is 67 because the numbers have been treated as strings. • Give students the opportunity to share their thoughts and then use the explanation provided (Pages 71 and 72). • Demonstrate the example code using int(). • Introduce (Page 53) Task B, 'I need an int'. Students will update their code to use int() and use it to perform calculations using arithmetic operators.	• Remind students about the colour coding of blocks and show them how to drag nested blocks in EduBlocks. • Offer assistance and ask individual students questions to ensure they understand why they need to add 'int()' to make their code work. (To turn the input from a text string such as "1" to an integer – whole number.)
3	• Introduce the data types table (Page 72). • Direct students to (Page 54) Task C, 'Using decimal numbers'. • Ask 'What happened when you ran the code? Why do you think that?' • Give students the opportunity to share their thoughts and then explain the error they see. • Talk through the demonstration on 'Use the float() function for decimal numbers' (Page 73). You could demonstrate this in EduBlocks if you prefer. • Direct students to complete (Page 54) Task D, 'Calculate ticket prices'.	• These are not all of the data types in Python, they will meet more later. • Reassure students that while Python errors look complicated, they will learn to understand what they mean. • Now they know that when they see 'invalid literal for int()' they should check the data type of the value that they are passing in. Add this to your class list of errors and solutions. • Reinforce that they must place the 'input()' statement inside an 'int()' or 'float()' if they want to use numbers. Encourage them to read the error messages and debug their code. When an input block is put inside an int block the value is converted to a number before it is stored.

Activity	Instruction	Guidance
4	• Introduce the material on comparison operators (📖 Page 74). • Compare to the similar blocks in Scratch. • Demonstrate creating the EduBlocks example in a new program in EduBlocks. Show students where the 'Logic' blocks are.	• Note that these blocks are all found under 'Logic' in EduBlocks.
5	• Direct students to complete (📓 Page 54) Task E, 'True or False?' to check their understanding of comparison operators.	• There is a tip in the Workbook for students to check the answers themselves using EduBlocks. This would be good for those that answer the questions quickly, though there may not be time to check all of them in this way. • Discuss the answers immediately. Students might have questions about Task E, question d. The answer here is True yet the total will be 10.0 and not 10. Python knows that these numbers are equal to one another, even though the data types are different. • Run the code to show that Python prints True or False. • Make the connection to the Boolean data type.
6	• Ask 'Why are data types important in Python? What data types do you know and what values can they have?' • Ask students to record their reflections in (📓 Page 55) 'Reflection'.	• Refer back to the data types table. • Data types tell Python how to treat data. • Sometimes you have to tell Python which data type to use, using int(), float(), str() or bool().
	Homework (Optional) (📓 Page 55) • Students match the Python blocks to their descriptions.	• Go through the answers with the students to check their understanding.

Build on this: **Make a decision**

Ask students to write a small program that asks the user for input and then makes a decision. For example, they could ask how old the user is and then tell them whether they are old enough to watch a particular film.

Chapter 5:	Lesson 3 Python: Boolean logic, comments and functions		
Student's Book page 75			
Learning Objective/s	• Add effective comments to a text-based program • Recognise patterns in algorithms and edit to use subroutines • Follow and debug algorithms that use subroutines		
New keywords	**Parameter** – Special named variable that is part of a subroutine definition		
You will need	• Student access to the EduBlocks website (Website C5L3R1) • Access to the starter project: (Website C5L3R2) • Access to the second starter project: (Website C5L3R3) • EduBlocks accounts for students (optional, but recommended)	**Key pedagogy**	Investigate

Activity	Instruction	Guidance
1	• Remind students of Boolean operators, AND, OR and NOT (Page 75). • Demonstrate running (not creating) the Book recommendation project. • (Website C5L3R4) • Share a link to the project and ask students to complete (Page 56) Task A, 'Decisions'.	• Relate to their knowledge of Boolean logic and AND/OR/NOT gates. • Note that these blocks are found in the 'Logic' section in EduBlocks. • Mention that the code uses '==' with the String type. • Observe students and check that they are confident with the meaning of the comparison operators. • Some students may recognise that they can use nested if statements or elif statements to improve the code. If they understand this and are able to then encourage them to refactor the code.
2	• Recap the purpose of comments and introduce comments in Python (Page 76). • Demonstrate adding a comment to the Book recommendation project. • Direct students to add comments to their projects in EduBlocks. • Direct students to answer (Page 56) Task B, 'Comments' when they finish.	• Students will need to clone the EduBlocks starter project before they can add comments. They can do this by clicking on the 'Clone' button in the top right-hand corner. • You can direct students to only add one or two comments, depending on time available.

Activity	Instruction	Guidance
3	• Ask 'Why is it important to add comments to your code?'	• It's often easier for a human to read a helpful explanation than to read the code. • You might forget why you wrote code the way you did. • Other coders can work on your code more easily.
4	• Explain that you can define your own functions in Python (Page 77). • Direct students to (Page 57) Task C, 'Recognising patterns'. • After giving your students some time to consider the questions in their Workbook, have a class discussion on the similarities and differences. The answers are provided in the answer sheet. • Use the explanation in the Student's Book to show how the three lines of code have now been used in a function instead.	• Explain that when we spot a pattern where some code is the same with only some small differences we should consider using a function. • Software engineers call this DRY – Don't Repeat Yourself.
5	• Open the second starter project showing the book recommendation app with a new function added. (Website C5L3R5) • The first two if statements have already been changed to incorporate the function which is the same process. • Demonstrate how to change the third if statement so that it calls the function instead of repeating the three lines of code. There is a walkthrough of this alongside the task in their Workbooks (Task D). • Ask students to complete (Pages 57 and 58) Task D, 'Modify the recommendation app'.	• Make sure you use a (" ") block for text and a ( ) block for variable names and numbers. This has also been pointed out in the Workbook as a tip.
6	• Remind students that they have already spent time debugging code in other chapters. They will now try to debug a function. • Instruct students to complete (Pages 58 and 59) Task E, 'Debugging a function call'.	• This is an error that they may have made themselves when they modified the recommendation app in the previous task.
7	• Ask the students 'Did you have any errors in the 'Modify the recommendation app' task? If so, how did you fix them?' • Ask the students to record their reflections in (Page 59) 'Reflection'.	• There will most likely be a few students that encountered errors with their code when they were modifying it. Keep a list of common bugs that they encountered.

Activity	Instruction	Guidance
	Homework (Optional) (Page 59) • Students are asked to think about the theme that they might decide to use for their recommendation app projects.	• There are lots of examples given in the book to support this. If students still need support then perhaps have a class discussion to generate more ideas.
	Build on this: **Add another book recommendation** Students could make up the name of a book that is about robots or elephants. They should decide which age group the book would apply to. Next, they should add another if statement that would display their recommendation.	

Chapter 5: Lesson 4 Start your recommendation project

Student's Book page 79

Learning Objective/s	• Create a text-based program that uses input, output, comparison operators and functions		
You will need	• Access to the starter project: (Website C5L4R1) • Access to the example projects: (Website C5L4R2) (Website C5L4R3) (Website C5L4R4) Student accounts required for saving work **or** students take screenshots of developed code to remake quickly at the start of the next lesson.	**Key pedagogy**	Develop practical skills

Activity	Instruction	Guidance
1	• Introduce the project brief and discuss the examples (Page 79). (Website C5L4R5) (Website C5L4R6) (Website C5L4R7) • Explain that students can refer to the complete example project (Page 80).	• Make sure students are clear on the requirements. • Explain that they will have two lessons to complete their projects.
2	• Talk through the example plan in the Workbook and ask students to complete (Page 60) Task A, 'Plan your recommendation system'.	• Make sure students are clear on the project brief and what they need to do. • Having a fully completed plan will make the coding go much more smoothly.

Activity	Instruction	Guidance
3	• When students are ready, provide them with the starter project. (Website C5L4R8) • Explain to the students that the starter code uses the variables 'age' and 'likes' to give them a structure for the code. They can create new variables based on their own plans and select their own variables using the drop-down menu. (This is also a tip in the Student's Book.) • There is also a tip to support them with debugging. The starter project has blocks of code that are disabled. This is to allow them to test their code as they go. To test that the block works, they should right-click and enable the code block.	• Provide debugging support as needed. Demonstrate enabling and disabling code blocks to support good testing. • Encourage students to help each other.
4	• Ask 'What errors have you found in your code and how did you fix them?' • Ask the students to record their reflections in (Page 61) 'Reflection'. • Make sure students save their projects as they will be continuing with them in the next lesson.	• Allow time for sharing problems so that students can learn from each other.
	Homework (Optional) (Page 61) • Students are asked to think about the problems that they encountered when starting their programs and how they overcame them.	• They may have been really fortunate and not experienced any issues at all. It is OK for them to reflect on this as well, and think about why that may have happened.
	Build on this: **Add additional inputs** Students could add additional user inputs and offer a wider choice of recommendations.	

Chapter 5:	Lesson 5 Complete your recommendation project		
Student's Book page 81			
Learning Objective/s	• Create a text-based program that uses input, output, comparison operators and functions		
You will need	• Saved projects **or** screenshots of code from previous lesson.	**Key pedagogy**	Develop practical skills

Activity	Instruction	Guidance
1	• Remind students of the project brief and objectives. • Talk through the steps they need to take (Page 81). • Direct students to (Page 62) Task A, 'Project checklist'. • They should complete the checklist as they go.	• You can encourage some students to go beyond the project brief and others to just focus on the core behaviour of the program.
2	• Ask 'How much progress have you made?' Give advice on how students can complete their projects. Encourage them to focus on making sure they have a project that can be used. • Direct students to (Page 62) Task B, 'Meeting the brief'.	• It is important to note here that the 'Project checklist' and the 'Meeting the brief' checklist are ongoing lists that they should check off throughout the project. • Observe students. Support with debugging and set stretch goals as needed.
3	• Direct students to (Page 63) Task C, 'Testing' when ready.	• If students need extra support here then you can remind them of the example project that they completed in Lesson 3, where they entered the numbers and characters that they wrote in order to display each recommendation. An example entry has also been provided in row 1 of the testing table.
4	• Ask 'What are you looking forward to about sharing your projects? Are you concerned about any parts of your project?' • Ask the students to record their reflections in (Page 63) 'Reflection'.	• Some students may be excited, but others may be nervous that there may be errors in their code. Reassure them that everyone is learning and that they will help each other to fix any problems they find.

Activity	Instruction	Guidance
	Homework (Optional) (Workbook Page 63) • This asks students to list three reasons why it is important to test their programs.	• Example responses might be: ○ to make sure the code works as expected ○ to make sure that the program is easy for a user to follow and use ○ Errors in code can frustrate the end user and make them not want to use the program again.

Build on this: **ASCII art (art using the characters on your keyboard)**

Students can use functions to create ASCII art with characters to decorate the output.

Chapter 5: Lesson 6 Share your recommendation project

Student's Book page 83			
Learning Objective/s	• Read and explain code written by another programmer		
You will need	• Projects to share	**Key pedagogy**	Showcase

Activity	Instruction	Guidance
1	• Remind students that they will be trying out each other's projects. • Ask them to check that they haven't included any personal information in their projects. • Show students how to share a project in EduBlocks, and tell them how you would like them to share projects (email, or just sitting together at a computer).	• You could put students in pairs to swap and review each other's projects.
2	• Organise students to give feedback on each other's projects. • Each student should write feedback in the Workbook of the student who created the project. Direct students to (Workbook Page 64) Task A, 'Feedback'. • Encourage students to be kind and helpful when they give feedback. They should use the structure in the Workbook to list two things that they liked and suggest one improvement.	• Observe and ensure that the feedback is constructive.

Activity	Instruction	Guidance
3	• If there is any time remaining then the students could read their feedback and make any improvements to their work.	• Mention that pair programming is popular in industry and it can be helpful to work together on a project.
4	• Instruct the students to complete the reflection activity in (Pages 64 and 65) Task B, 'Reflection'.	
5	• Ask 'When you tested a partners program, what did you think of the recommendations you were given? Do you think you will follow the recommendations?' • Celebrate what they have achieved: writing a Python program that gives recommendations to someone else.	• Make the connection to real-world recommendation systems that are based on data. • Make the connection back to machine learning that can make recommendations based on lots of data.
6	• Reflection: In pairs or small groups, ask students to discuss 'How did you enjoy creating a recommendation system that could be used by others?' and then complete (Page 65) 'Reflection'. • Students can then complete the (Page 65) Project skills checklist.	
	Homework (Optional) (Page 65) • Students are asked to imagine what their programs would look like if they had made them in Scratch instead. They should draw an outline using the space provided.	• Link back to their use of Scratch in Chapters 1 and 3.

Build on this: **Text-based Python**

The students now have a completed program in EduBlocks. They should spend some time exploring the text-based version of their program on the right-hand side. What are the similarities? What are the differences?

Chapter 6 – Connect the world

Project: Create a method for encrypting text

Chapter overview

In this chapter, students will recall IP addresses and learn how these relate to URLs. They will be introduced to DNSs and reminded about secure websites. They will then discover different types of wireless transmission and discover the common errors that might occur for each type. Next, they will work in pairs to develop their own encryption algorithms, use them to write messages and then pass these on to another pair to crack.

Chapter 6: Lesson summary		
Lesson	**Learning Objectives**	**Summary**
1	• Describe the function of an IP address and a URL • Explain the role of DNS in accessing websites • Explain how to identify secure websites	Students will recall their knowledge of IP addresses and URLs and identify the parts of a URL including domain names. They will explain the way the DNS maps from a domain name to an IP address. Finally, they will learn how to check whether a website is secure.
2	• Compare Bluetooth, Wi-Fi and cellular networks, including the different generations • State the common reasons why errors can occur in data transmission	Students will recall their knowledge of wireless communication and compare the features of Bluetooth, Wi-Fi and cellular networks. They will recognise cellular masts. Finally they will state the reasons why errors can occur in data transmission.
3	• Define encryption and explain its role in keeping data secure during transmission	Students will recall their knowledge of the Pigpen and Caesar cipher encryption algorithms. They will recognise these as substitution ciphers and will encrypt and decrypt messages using an existing encryption algorithm.
4	• Design an encryption algorithm	Students will be introduced to the project brief where they will design their own substitution cipher. In pairs, students will design an alphabet substitution cipher where each letter of the alphabet is replaced by a symbol. They will begin creating Scratch sprite costumes for each letter.
5	• Design an encryption algorithm	Pairs will complete the design of their algorithm and encode a message using their encryption algorithm. They will use the terms 'decryption algorithm' and 'key' and prepare clues for use in cracking their code.

6	• Crack an encryption algorithm	Pairs will attempt to crack a message encoded with another pair's encryption algorithm. Clues, related to real-world decryption scenarios, will be provided as needed. Students will relate their experience to real-world encryption.

End of chapter project	**Create a method for encrypting text**
Example ideas that students could design:	Students will create an encryption algorithm that uses a substitution cipher that replaces letters of the alphabet with unique symbols. Examples include: • assigned emoji characters to letters of the alphabet • using the braille alphabet • using sign language fingerspelling letter images • drawing 26 unique symbols • using sprite costumes in 26 different positions.

Why we are learning this	
Useful skills covered in this chapter	• The role DNS plays when they access websites. The features of different wireless communication methods that they use in their everyday life. Real-world examples and practical experience of encryption and decryption algorithms to create and crack codes.
Links to future careers	• Cryptographer • Software engineer
Links to other subjects	• History

Chapter 6 – Connect the world

Project: Create a method for encrypting text

Name	

	Working towards (1)	**Meets expected (2)**	**Exceeding expected (3)**	**Score**
Created a substitution cipher	• Partially created a substitution cipher	• Created a substitution cipher and encoded a message with it	• Created a substitution cipher and encoded a message with it. Demonstrates an understanding of how difficult it will be to crack the code.	
Cracked a substitution cipher	• Attempted to crack a substitution cipher	• Cracked a substitution cipher, using a clue as appropriate	• Cracked multiple substitution ciphers with increasing speed	
Uses appropriate vocabulary	• Uses some appropriate vocabulary	• Describes encryption and description and substitution ciphers	• Describes encryption and description and substitution ciphers. Uses additional vocabulary such as frequency analysis.	

Teacher feedback	
Learner response	

Chapter 6:	Lesson 1 Internet standards: DNS and HTTPS		
Student's Book page 86			
Learning Objective/s	• Describe the function of an IP address and a URL • Explain the role of DNS in accessing websites • Explain how to identify secure websites		
New keywords	**Domain name** – Human-readable label that is assigned to an IP address **Hypertext Transfer Protocol Secure (HTTPS)** – Secure way of sending data between a web browser and a website **Top-level domain (TLD)** – The last part of the domain. For example .com **Domain Name System (DNS)** – Naming system and way of mapping from human-readable domain names to IP addresses		
You will need	• No special equipment	**Key pedagogy**	Classroom talk

Activity	Instruction	Guidance
1	• Remind students that web pages have addresses called URLs (📖 Page 86). Describe the parts of a URL and introduce the definitions for scheme and domain name.	• URL stands for Uniform Resource Locator.
2	• Explain that it's important to know whether a website is secure (📖 Pages 86 and 87). • Demonstrate how to check whether a website uses HTTP or HTTPS, using the web browser that your students have access to. • Direct students to complete (📓 Page 66) Task A, 'Identify the parts of a URL'.	• Note that web browsers used to use a locked/unlocked padlock to indicate whether a website is secure. This is changing to use a settings icon that you can click to view more information. This reflects HTTPS becoming the default for websites rather than something exceptional. • Give students guidance on what websites they are allowed to visit. You could direct them to specific websites or ask them to choose sensibly.

Activity	Instruction	Guidance
3	• Introduce the material on domain names (Page 87). • Discuss the top-level domain for your country. • Ask students to complete (Page 66) Task B, 'Top-level domains'.	• Cisco are a provider of networking solutions and they own the .cisco TLD. • Does your country just have a two-letter code such as '.fr' or does it use two-part TLDs such as 'co.uk'? ('.uk' is also a TLD that can be used with a main domain.) • If some of your students are from other countries, then you could discuss their TLDs too. • Domain names are sometimes bought and sold. Some students may have heard about buying domain names. This is done through a domain registrar who registers the name with a domain registry.

4	• Introduce the material on the Domain Name System (DNS) including talking through the steps in the diagram (Pages 88 and 89).
	• In pairs, ask students to take turns explaining the steps to each other. Ask them to discuss 'What role does the Domain Name System play in accessing a website from a web browser?'
	• Ask students to complete (Page 66) Task C, 'The Domain Name System'.
		• Briefly mention that web browsers and DNS servers can 'cache' or store results for quicker lookups. But they can only store results for a limited time as IP addresses can change.
		• Emphasise that DNS lookups on the internet are very fast, and the distributed DNS means that a single computer doesn't need to store all of the IP addresses – there are a lot!
		• Task C:
		  ◦ The Domain Name System maps from human-readable names to numbers that computers can work with.
		  ◦ The DNS means that a single computer doesn't need to store all of the addresses – the work is distributed.

Activity	Instruction	Guidance
5	• Introduce the material on DNS security (📖 Page 89). • Reflection: In pairs or small groups, ask students to discuss 'Why is it important to have standards such as IP, DNS and HTTPS?' and then complete (📓 Page 67) 'Reflection'.	• DoH is a new technology but may get adopted rapidly as HTTPS did. • These standards are needed so that computers on the internet are able to communicate reliably and securely. • Imagine if everyone wanted to use a different naming convention or different security protocols. Clients and servers would not be able to talk to each other.
	Homework (Optional) (📓 Page 67) • As students travel around their community during the week, they should look out for TLDs on signs, buildings, vehicles, adverts and so on, and make a note of any that they see.	• Advise students to look for patterns such as common TLDs in their community or infrequent use cases. • You could ask students what they found out as an introduction to the next lesson.

Build on this: **Research the history of internet standards**
Students could discover the role of these individuals:
Joyce K Reynolds, Paul Mockapetris, Elizabeth Feinler, Nabil Bukhalid, Alissa Cooper

Guidance:
Joyce K Reynolds – significant contributor to early internet standards
Paul Mockapetris – created the Domain Name System
Elizabeth Feinler – managed the list of mappings from domain names to internet addresses before DNS. There weren't many computers on the internet at first!
Nabil Bukhalid – established the Lebanese domain registry
Alissa Cooper – has a leadership role in internet standards including HTTPS

Chapter 6: Lesson 2 Wireless data transmission

Student's Book page 90

Learning Objective/s	• Compare Bluetooth, Wi-Fi and cellular networks, including the different generations • State the common reasons why errors can occur in data transmission		
New keywords	**Bluetooth** – Wireless standard for communicating between nearby devices **4G and 5G** – Versions of cellular network technology with improvements over previous generations **Packet loss** – When 'packets' of data fail to reach their destination **Bit errors** – When data gets changed during transmission		
You will need	• It's useful if you know the locations of local cell towers or masts.	**Key pedagogy**	Real-world contexts

Activity	Instruction	Guidance
1	• Remind students that digital devices can transfer data wirelessly using radio waves. Discuss the differences between Bluetooth, Wi-Fi and Cellular transmission (Page 90). • Introduce (Page 68) Task A, 'Which wireless technology?' and give students time to write their answers.	• If you have micro:bit devices, you could show one and explain that it can be programmed using Bluetooth and also has its own way of communicating between devices using radio waves. • There is some room for debate on the answers. The first example would typically be Bluetooth, but you could imagine headphones that are directly connected to the internet with Wi-Fi to stream music. • Observe as individuals or groups complete the Workbook activity and bring any interesting points into the following discussion.
2	• Ask 'Why do we have multiple wireless communication technologies? Couldn't we just have one?'	• They have evolved over time for different uses and have different features. Cost of manufacture and speed and range of communication needed are important factors. • Maybe in the future there will be a new technology that can do everything well.
3	• Introduce cellular networks (Page 91). Connect this to local knowledge. Do students know where the local cell towers are? Do they notice when they switch between cell towers? • Ask 'Why do you think mobile providers share cell towers?'	• Capture the shared knowledge of the class of cellular (mobile) cell towers or roof-top masts in your area. Does anyone know which providers they support? Some masts may have base stations for multiple mobile carriers. • Some students may recognise that their network provider has better coverage in some areas than others. • Who are the major mobile phone providers in your region? • Mobile providers share cell towers due to cost and reducing impact on the environment. In some countries there are laws that require new masts to be shared.
4	• Compare 4G and 5G cellular network standards (Page 91). • Introduce (Page 68) Task B, '5G features'.	• Link to the current adoption of cellular technology in your country at this time. • Has 5G been widely adopted?

Activity	Instruction	Guidance

5	• Explain how errors can occur in wireless transmission and the differences for different technologies (📖 Page 92).
	• Ask 'How can you reduce the chance of errors when using wireless transmission?'
		• You could:
		  • avoid physical obstacles
		  • move away from devices that could cause interference
		  • wait for a less busy time
		  • move closer to the Bluetooth device you want to communicate with (or wireless access point or cell tower).

6	• Ask students to complete (📒 Page 69) 'Reflection' and discuss their responses.	• Students may have noticed that their Bluetooth devices become disconnected if they move away from their phone. • They may get a better Wi-Fi connection in some places than others. • They may notice that the mobile phone cellular signal is better in some places. • They might imagine that 6G will be affordable for everyone and ultrafast!
	Homework (Optional) (📒 Page 69) • During the week, students should move around with their wireless devices and observe how position affects the connectivity.	• If they do not have wireless devices, then they can ask for permission to carefully use the devices of a trusted adult. • You could ask students what they found out as an introduction to the next lesson.

Build on this: Design 6G

Ask students to research plans for 6G and consider what they would want to add to the design. This could include features to lower cost, improve support for the Internet of Things or reduce the impact on the environment.

Chapter 6: Lesson 3 Encryption and data security

Student's Book page 93

Learning Objective/s	• Define encryption and explain its role in keeping data secure during transmission		
New keywords	**Encryption algorithm** – Specific set of rules for turning a plaintext message into encrypted ciphertext, sometimes with the use of a key **Decryption algorithm** – Specific set of rules for turning a ciphertext message into a plaintext message, sometimes with the use of a key **Key** – Piece of information that can be used with a decryption algorithm to decrypt a ciphertext message **Substitution cipher** – Way of encoding messages where a letter is replaced by another letter or symbol		
You will need	• Students will need access to computers that can run Scratch. • The encryption starter project <u>(Website C6L3R1)</u>	**Key pedagogy**	Classroom talk

Activity	Instruction	Guidance
1	• Remind students of the definition of encryption and introduce encryption algorithm, decryption algorithm and key (Page 93). • Discuss the Caesar and Pigpen ciphers (Pages 93 and 94). • Ask students to complete (Page 70) Task A, 'Decryption'. They will need to try to remember the grid for Pigpen.	• Emphasise that the keys for Caesar and Pigpen are designed to be easy to remember so they don't need to be written down. • These ciphers would have prevented casual observers from reading the text. • You could run a short timed activity as a class to give students a set amount of time to try to memorise the grid for Pigpen before they complete question 2. • The Caesar Cipher message plain text is: REMEMBER DO NOT REPLY TO TROLLS BECAUSE IT JUST MAKES THEM WORSE • The PigPen message plaintext is: MAKE SURE YOU ALWAYS CHOOSE A STRONG PASSWORD TO KEEP YOUR INFORMATION SAFE
2	• Introduce the material on data security (Page 94). • Ask students to complete (Page 70) Task B, 'Encryption'.	• Recall coverage of HTTPS and web browser security from Lesson 1 of this Chapter, and link these to the use of encryption.

Activity	Instruction	Guidance
3	• Introduce substitution ciphers and Atbash (Page 95). • Ask 'What is the decryption algorithm for Atbash? How easy would it be to decrypt a message coded using Atbash? How does it compare with the Pigpen and Caesar ciphers?' • Demonstrate encrypting a short ciphertext message of your choice on the whiteboard.	• If culturally relevant to your students, draw attention to the fact that one of the earliest encryption algorithms was for the Hebrew alphabet. • The decryption algorithm for Atbash is exactly the same as the encryption algorithm.
4	• Open the Scratch project (Website C6L3R2). • Click 'See inside' and then on the Stage code area. Demonstrate selecting the different messages in the broadcast block (for the different encryption algorithms) and clicking the green flag to encrypt the message. • Allocate each student a partner with whom they will swap messages, or ask them to find a partner when they are ready. It is important that they do not see each other's messages so they should not be seated close to each other. • Show students how to open the Scratch project on their computers. • Tell students that they can choose the encryption algorithm and key and that they are going to write the encrypted message in their partner's Workbook for them to decode. • Direct them to 'Encrypt a message' (Pages 95 and 96) and the corresponding task (Page 70) Task C, 'Decrypt a message'.	• Mention that there are space characters between words. Some words will wrap onto the next line. A new word only starts if there is a space. • Make sure that students know that they must not write their message in their Workbook. They should just type it in the Scratch project. • Suggest that they choose a line from a book or song, or a fact about a place, an animal or a person. They should not include personal information about themselves in the message.
5	• Students who correctly decrypt the message can move on to (Page 71) Task D, 'Explore the encryption algorithm'.	• It's fine if not all students have time to complete Task D. • If you have time, you can walk through the encryption algorithm as a class.
6	• Reflection: In pairs or small groups, ask students to discuss 'How much more difficult would it be to decrypt a message if you didn't know the encryption algorithm and have the key?' and then complete (Page 71) 'Reflection'.	• It would be much more difficult. • They will be learning more about this in the next lesson.

Activity	Instruction	Guidance
	Homework (Optional) (📖✏ Page 71) • This asks students to create a message and encrypt it using Atbash. They then need to explain the decryption algorithm for Atbash to someone else and ask them to decipher their message. Finally, they reflect on whether they explained it well enough so that their partner was able to decipher the message.	• The message should be something short. This task focuses on the ability of the student to explain the decryption algorithm rather than the skills of the other person to decipher a really long message. • The other person should be someone who wasn't in this class, ideally someone at home.

Build on this: **Try different ciphers**

Students could try encrypting more messages with the different ciphers in the Scratch project and ask a partner to decrypt them.

Chapter 6: Lesson 4 Create your own encryption algorithm

Student's Book page 97

Learning Objective/s	• Design an encryption algorithm		
You will need	• The encryption starter project (Website C6L4R1)	**Key pedagogy**	Develop practical skills

Activity	Instruction	Guidance

1	• Introduce the project brief and talk through the example ciphertext messages (Page 97).
	• Open the Scratch project that they will use to encrypt their messages. (Website C6L4R2)
	• Explain how the project works:
	  ◦ Click on the code area for the Stage and explain that the code sets a message and then broadcasts the message to choose the encryption algorithm.
	  ◦ Tell students that the sprite for the chosen algorithm will then encrypt the message and show it on the Stage.
	  ◦ Demonstrate the emoji, AAA and shapes ciphers by changing the message via the drop-down in the broadcast block and then clicking the green flag. You can also change the set message variable block.
	  ◦ There is a sprite called 'My cipher' which has 26 costumes each with a '?'. Students will need to replace each '?' with a unique character or design.
	• Ask students to plan their ciphers in pairs and complete (Page 72) Task A, 'Plan your substitution cipher'.
		• The code for the encryption algorithm is provided and is the same as the code for the Pigpen algorithm – each letter of the alphabet corresponds to a sprite costume.
		• Observe students planning their ciphers and make sure that their ideas are feasible to create in Scratch in two lessons.
		• The simplest ciphers will be ones that just swap letters or substitute emoji characters.
		• Costumes can also be created by drawing with the Paint editor or taking photos with the camera.
		• The starter Scratch project is set up to use messages created with letters from the English alphabet. You could amend this starter project to use a language with a different character set by adding/removing and renaming costumes. If you are doing this you should amend and test your new starter project in advance then use it in the demo to students.

2	• Remind students that they need to update the 'My cipher' sprite costumes with their own symbols. • Provide students with the link to the Scratch starter project and tell them how they should save their work. • Demonstrate how to enter emoji characters on the school computers. Chrome web browser: right-click -> Emoji Windows: Windows Key + '.' Mac: Ctrl + Command + Space • Ask students to start creating the 26 sprite costumes for their cipher by completing (Page 72) Task B, 'Start creating your substitution cipher'.	• Do the students have Scratch accounts or will they need to save their projects to the local network? They will continue with the project in the next lesson, so it is very important that they know how to save and load their projects. • Emoji characters can also be copied from an emoji website. The emoji Characters sprite also has a costume with emojis as text that can be copied and pasted using Ctrl+C/Ctrl+V. • Make sure that students use the sprite provided or ensure that their sprite costumes are labelled A–Z plus one additional costume with the space ' ' character as its name.

Activity	Instruction	Guidance
3	• Give the students regular time checks throughout the lesson so that they can adjust the complexity of their planned cipher if needed.	• Students can click the green flag at any time to check their progress with the sample message.
4	• Ask students to save their work and make notes to remind them what they need to complete in the next lesson. They should complete (Page 73) Task C, 'Save and load your Scratch project'.	• Make sure that projects are saved as students will be continuing with them next week.
5	• Explain that, in the next lesson, students will be completing their ciphers and encrypting a message.	• You could ask them to begin thinking about what their message will be. You can introduce this at any point in the lesson if students need a break from creating sprite costumes.
6	• Reflection: In pairs or small groups, ask students to discuss 'Did you overcome any issues with your Scratch project? If so, share them' and then complete (Page 73) 'Reflection'.	
	Homework (Optional) (Page 73) • This asks students to have a conversation with people they know about coded messages. Have they ever communicated in coded messages? They should discuss why they decided to do it, the rules and methods used, and whether it was successful.	• Students should ask a range of people to find someone who has used coded messages. These experiences are likely to be from childhood. • If you have used coded messages, you could share the details at the start of the next lesson to start a short discussion.

Build on this: **Code the encryption algorithm**

You could ask confident coders to code the encryption algorithm themselves, using their Workbook notes from the previous lesson if necessary.

Chapter 6	Lesson 5 Complete your encryption algorithm		
Student's Book page 99			
Learning Objective/s	• Design an encryption algorithm		
New keywords	**Crack** – To find out how to get the plaintext from encrypted ciphertext without knowledge of the decryption algorithm or key **Frequency analysis** – Knowledge about how often each letter is used in text written in a particular language		
You will need	• Students will need their saved Scratch encryption projects from the previous lesson. • You will ideally need access to a colour printer so that completed codes can be printed ready to be shared next week. Alternatively, you can save images of the code and share them digitally. • Make sure you are familiar with saving an image of the Scratch stage (right-click) and then opening the file and printing it on your school computers.	**Key pedagogy**	Develop practical skills

Activity	Instruction	Guidance
1	• Introduce the material on 'Cracking a code' (📖 Page 99).	• Gauge how much time students need to create their codes. You could allocate the material in the Student's Book for self-study if there is not enough time to cover it during the lesson.
2	• Introduce the material on frequency analysis. (📖 Pages 99 and 100).	• Have your students heard of al-Kindi before? Can you make any culturally relevant connections to this ancient thinker?
3	• Direct students to complete their codes and then move on to (📖 Page 74) Task A, 'Prepare a message' and (📖 Page 74) Task B, 'Prepare a clue' to prepare a message and clues. • Show students how to set the 'message' variable in the code for the Stage.	• Check that students have chosen a clue that is 5–7 characters long so that it will provide several letters to the pair decrypting their message.

Activity	Instruction	Guidance
4	• The remainder of the lesson is allocated to completing their code. • Remind students to save their projects.	• Encourage students to work more quickly, if necessary. • Check that students have added their own message and filled in their team name on the backdrop (this will help to identify the codes in the next lesson if clues are needed).
5	• Explain to students how they will be sharing their ciphertext messages with other pairs. If you have access to a colour printer, you will gather the prints from students this lesson and distribute them in the next lesson so that each pair has another pair's message. (If you do not have access to a printer you will need to save the images to a shared location and then assign them to pairs – you could indicate the pairs in the file names.)	• You may want to keep the printed codes to prevent them being lost before the next lesson. Names should be written on them for identification. • If you will be printing the ciphers between lessons, make sure you have copies of the images. Students could save them on the local network or email them to you. • It would be helpful to have multiple copies to give to early finishers or those who need an additional copy.
6	• Ask students to complete (Page 74) Task C, 'Cracking your code'.	• This is a reflection activity that students can complete while you are ensuring that copies of codes are available for the next lesson.
7	• Ask 'Have you ever cracked a code without knowing the encryption algorithm or key? How did you do it?' • Ask students to record their reflections in (Page 75) 'Reflection'.	• Students may have played escape games or cracked codes in puzzle books. They may have been introduced to code cracking in a coding club or school activity.
	Homework (Optional) (Page 75) • This asks students to create a piece of artwork that shows their name in their code.	• Give students time before they leave the lesson to write or draw the symbols from their code that represent each letter in their name. • The artwork can be used as an extra clue in the next lesson when teams are solving the code.

Build on this: **Frequency analysis**

Photocopy the same page of text from a book and give it out to students. Allocate each student one or more letters from the alphabet and get them to count the number of occurrences.

Compile a frequency table and use it to work out the percentages of each letter. Compare it with the frequency analysis table on page 101 of the Student's Book.

Chapter 6:	Lesson 6 Cracking the code		
Student's Book page 101			
Learning Objective/s	• Crack an encryption algorithm		
You will need	• Printouts or digital copies of encrypted messages and clues from the previous lesson. • It would be helpful to have multiple copies to give to early finishers or those who need an additional copy.	**Key pedagogy**	Showcase

Activity	Instruction	Guidance
1	• Explain that during this lesson students will be cracking the codes prepared by other pairs (Page 101). • Talk through the section on letter frequency in English, and tell students this knowledge will be useful when they crack the codes. • Hand out printed codes to pairs. • There is space in the Workbook for notes (Page 76) Task A, 'Crack the code', recording the clue (Page 76) Task B, 'Clue' and the plaintext message (Page 77) Task C, 'Plaintext message'. • A letter frequency table is provided (Page 101).	• Encourage students to make notes on their printed ciphertext sheets and/or in their Workbooks. • Note that words wrap around onto the next line. There is a space character between words.

2	• Assess how students are progressing and decide when clues should be given.	• Additional tips that can be given:
		  • Direct students to the letter frequency table and identify the most common letters.
		  • Encourage them to think about what one and two letter words could be.
		  • Do they have any double letters in their message? Which letters can be repeated in English?
		• Clues are of the form:
		Word number N is ______.
		• Clues were recorded in Task B.

Activity	Instruction	Guidance
3	• Ask early finishers to try to crack the example codes (Page 97). • Give additional copies of codes to early finishers. • If any pairs are struggling after they have been given clues, then they can be shown the key in the Scratch project.	• Scratch projects could be shared publicly if students have accounts. Make sure they have not included personal information in their messages. Also, make sure students do not open the Scratch project, click on the Stage code area to view the original message and copy the answer without cracking the code. • The costumes for the sprite provide the key. • It is also possible to add a 'broadcast [plaintext]' block to print the plaintext message under the ciphertext message. • The message from the emoji, shapes and AAA Student's Book examples is 'make sure you always choose a strong password to keep your information safe'.
4	• Ask students to complete (Page 77) Task D, 'I cracked a code!'. • Ask students to complete (Pages 77 and 78) Task E, 'Reflection'.	• Students should correctly use the vocabulary: 'encryption algorithm', 'decryption algorithm', 'key', 'plaintext' and 'ciphertext'. They may also use: 'cipher', 'frequency analysis', 'clue' and 'secure/security'.
5	• Ask 'What helped you to crack the code? What made it more difficult?' • Celebrate the achievements of students. • Ask student to imagine what it would have been like to be al-Kindi cracking substitution ciphers for the first time. • Explain that modern encryption algorithms are much more complex but the principles are the same.	• Students may find it harder to decrypt messages that use symbols that are similar. • Messages that include unusual letters or combinations of letters are harder to decrypt. • Longer messages have more letters to give students clues and do frequency analysis. • How would a computer solve codes created with images? Link back to image detection in machine learning from Chapter 1.

Activity	Instruction	Guidance
6	• Reflection: In pairs or small groups, ask students to discuss 'Did you enjoy cracking the codes? Explain your answer' and then complete (Page 78) 'Reflection'. • Students can then complete the (Page 78) Project skills checklist.	
	Homework (Optional) (Page 78) • Give students a second printout of their code for them to take home. They should give it to someone else to solve. Did they solve it? How many clues were needed?	• Students should explain frequency analysis to the person they give the code to so that they have a starting point for solving the code.

Build on this: **Crack more codes**

Print out more copies of the codes and ask each pair to crack more codes. Do they get faster as they get more familiar with the techniques? Are some codes easier to crack than others? Why? Which codes could be hidden in plain sight without anyone even realising there is a code?

Chapter 7 – The power of data

Project: Help the school to make a student-led decision about where to spend some money

Chapter overview

In this chapter, students will create a model to simulate a royalties calculator and then use the model to make data-driven decisions. They will import their data into Scratch to see the data visualised in a different way. Students will explore primary keys and relate data across tables. They will add conditional formatting to help to analyse data and understand how to search data in a database. Students will evaluate and create data capture forms that link to spreadsheets and use their new skills to survey students at school to make a student-led decision about where to spend money.

Chapter 7: Lesson summary		
Lesson	**Learning Objectives**	**Summary**
1	• Identify systems that model real-life scenarios • Identify simulators in industry that model real-life scenarios	Students will research the use of simulators in industry and understand the importance of using accurate data to make data-driven decisions. They will create a spreadsheet to model two scenarios to answer questions. Students will import their data into a Scratch project to see the data represented graphically.
2	• Define a primary key and describe its purpose • Select an appropriate field to be a primary key • Write rules to apply basic conditional formatting to cells • Write rules to apply complex conditional formatting to cells	Students will practise identifying and suggesting primary keys. They will use a primary key to join data to help to answer questions. They will apply single colour and colour scale conditional formatting to assist in data analysis.
3	• Describe how to search databases using single criterion, such as <,> • Evaluate the effectiveness of using data capture forms	Students will identify which results will be returned for single criterion database queries. They will look at uses of data capture forms and tips for designing effective forms. Students will evaluate example data capture forms then create their own data capture forms linked to spreadsheets. They will preview and test their forms to look at them from a user's perspective.
4	• Design and create a data capture form • Capture and store data	Students will be introduced to the project for this chapter. They will design their forms by thinking about the questions they want to ask and the best method to use for asking them. They will start to capture data from other students to populate a spreadsheet.

5	• Use conditional formatting to help to analyse data and make decisions	Students will look at student examples of conditional formatting based on a similar project. They will use their captured data and apply conditional formatting to help them to reach conclusions. Students will summarise their findings and their process and then practise delivering their presentations ready for the following lesson.
6	• Present your analysis and decision • Evaluate the effectiveness of your data capture	Students will present their findings and the process they used to reach the decision. They will complete a reflection activity to evaluate how successful their forms and spreadsheets were at answering the project brief.

End of chapter project	**Help the school to make a student-led decision about where to spend some money**
Example ideas that students could design:	The students will design a data capture form to gather ideas from students. They will analyse the results and use conditional formatting to help them to make a student-led decision about how to spend some money. They will present their analysis and final decision to the class. Their project will include: • a Google data capture form • a Google sheet with conditional formatting • a showcase of their analysis and decision.

Why we are learning this	
Useful skills covered in this chapter	• Students may need to design data capture forms for others to fill in • They can evaluate the effectiveness of forms they are asked to fill in at school or outside school
Links to future careers	• Data capture forms are used widely in many careers for surveys, competitions, feedback and more
Links to other subjects	• Business – gathering feedback on project ideas

Chapter 7 – The power of data

Project: Help the school to make a student-led decision about where to spend some money

Name	

	Working towards (1)	Meets expected (2)	Exceeding expected (3)	Score
Input from students	• Information from the data capture form is either from a limited audience or does not feed to a linked spreadsheet	• A spreadsheet of information represents an ideal audience size and has been fed from a data capture form	• A spreadsheet of information represents an ideal audience size and has been fed from an optimally designed data capture form	
Data driven analysis	• A decision has been made without analysis of the data or use of conditional formatting	• A spreadsheet shows conditional formatting to visualise the data driven decision	• A conditionally formatted spreadsheet with additional formatting or formulae that enhances visualisation of the data driven decision	
A data driven showcase	• There is no clear evidence presented as to why the decision was made	• The results are presented and it is clear how the data driven decision was made	• The results are presented visually using the spreadsheet as a prop. It is clear how the data driven decision was made	

Teacher feedback	
Learner response	

Chapter 7: Lesson 1 Modelling data to make decisions

Student's Book page 104

Learning Objective/s	• Identify systems that model real-life scenarios • Identify simulators in industry that model real-life scenarios		
New keywords	**Simulator** – App or machine that realistically models real-world processes and conditions		
You will need	• Computers with access to a search engine and spreadsheet software such as Google sheets • The link to the Scratch starter project (Website C7L1R1)	**Key pedagogy**	Real-world contexts

Activity	Instruction	Guidance
1	• Go through what students already know and define the new keyword 'simulator'. • Talk through examples of simulator games (Page 104). Ask 'Which simulator games have you seen or played?'	• They have already used spreadsheets in previous stages. In Stage 6, students investigated the use of data in a range of industries. • During Chapter 1 of Stage 7, students used a machine learning program that simulated the impact of intersecting asteroids. • Reference any simulator games that are trending locally. • Often the official websites for simulator games have trailers. If the students are unfamiliar with simulator games you could play a trailer. Remember to check the age rating of the game is appropriate for your students.

Activity	Instruction	Guidance

2	• Talk through the examples of simulator use in industry, highlighting the wide variety of careers that make use of simulators (Page 105).
	• Ask what are the disadvantages of using a training simulator?
	• Ask students to complete (Page 79) Task A, 'Simulators in industry'.
	• Ask 'Why is accuracy of data important in simulations?'
		• Students might give the following answers to the second discussion question:
		  • You might not have the same emotions or stress level when using a simulator.
		  • You know that the simulator can be restarted.
		  • The environment might be more complex than the simulator is able to mimic.
		  • The initial costs are so high that the simulator will need to have been in use for a long time before cost benefits can be seen.
		• Students should note their sources when doing the Workbook activity. Remind students of the plagiarism and citation work from Chapter 2 of this Stage.
		• There is a lot to get through in this lesson so limit the time for Task A. It should be a short research activity into one use of a simulator in industry.

3	• Introduce spreadsheets as a data modelling tool to simulate real-world scenarios. • Read through the radio station brief (Page 106) with the class. • Get students set up on computers with their spreadsheet apps.	• If students are using an online spreadsheet tool, they could collaborate to build on their skills from Chapter 2 Content creation. • The Student's Book doesn't mention currency as a specific currency is not important to the task. When reading through the content you can describe the amounts in your local currency.

Activity	Instruction	Guidance
4	• Students work through the Build tasks (📖 Pages 106 and 107). • Students should complete the questions in (📓 Pages 79 and 80) Task B, 'Data-driven decisions' after carrying out each of the Build tasks.	• The Build sections are in three parts; Radio Station 1, Radio Station 2 and Data Visualisation. At the end of each section, there is a Workbook activity with questions to answer, based on the information they have at that stage. • Students must save their spreadsheets as they will be using these again in the next lesson. • During the 'Data Visualisation' build, students will need access to the Scratch project. They do not need to save this project as it will not be revisited in future lessons. • There are plenty of options to extend the activity, for example, by formatting the spreadsheet, changing the Scratch visualisation and experimenting with different royalties and targets. • Alternatively, students could design a question for another student to answer such as: 'If the singer wanted to go on holiday in 20 days, what royalty should they negotiate?' In a spreadsheet this is called "goal, seek".
5	• Reflection: In pairs or small groups, ask students to discuss 'Did you prefer seeing the data in the spreadsheet or represented graphically in the Scratch project? Why?' and then complete (📓 Page 80) 'Reflection'.	• This is a personal preference. Some students may love seeing all the data on one page in a spreadsheet and enjoy the simplicity in adjusting data and calculations. Others may prefer the visual interface of Scratch and the ability to customise it to tell a story.
	Homework (Optional) (📓 Page 80) • Students are asked to think of an idea for a computer game that is based on a simulation. They should describe the game and the data that it would use.	• There are examples of simulation-based computer games (📖 Page 104). • Example: a zoo simulator would have data on the number of visitors and the cost of food for the animals.

Activity	Instruction	Guidance
	Build on this: If you have access to education simulators at your school, you could get students to use them. Examples might be chemistry lab simulations or business studies forecasting simulations. Ask students to give feedback about their experience and use their new knowledge to explain the advantages/disadvantages of using simulations.	

Chapter 7:	Lesson 2 Organising and formatting data		
Student's Book page 110			
Learning Objective/s	• Define a primary key and describe its purpose • Select an appropriate field to be a primary key • Write rules to apply basic conditional formatting to cells • Write rules to apply complex conditional formatting to cells		
New keywords	**Record** – Row in a database table **Field** – Column in a database table **Primary key** – Assigned field in a database table that contains data that is both unique and not empty for every record **Conditional formatting** – Rule with criteria that, if met, changes the appearance of cells within a spreadsheet		
You will need	• Computers with access to spreadsheet software such as Google sheets • Students' royalty modelling spreadsheets from Lesson 1	**Key pedagogy**	Develop practical skills

Activity	Instruction	Guidance
1	• Go through what students already know and define new keywords 'database' and 'primary key'.	• In the last lesson, students used data in a spreadsheet, added calculations and reached decisions. Students have been using colour to format text during their Content creation lessons since Stage 2.

Activity	Instruction	Guidance
2	• Read through the organising data examples to introduce primary keys as those containing unique, non-null values (Pages 110 and 111). • Relate back to the data used in the last lesson and ask what field would make the best primary key. • Ask students to complete (Page 81) Task A, 'Identify primary keys'.	• With the example from the last lesson, all fields are unique and non-null. The purpose here is to think about future proofing the model. There is potential that the song can be played the same number of times and therefore earn the same amount. There is also potential that the song is not played at all one day so the total amount would remain the same. The only field that could be used is 'Date'.
3	• Introduce conditional formatting with the examples for single colour and colour scale formats (Page 111). • Remind students of the scenario from the last lesson where the singer selected radio station 1 to play their song based on the data model they created. • Students should work through the Build tasks using their spreadsheets from last lesson as the starting point (Pages 112, 113 and 114). • Ask 'Did you find a relationship between the DJ and the times played?' and 'Why was linking tables and applying conditional formatting useful for visualising the relationship?'.	• The Student's Book screenshot examples use Google Sheets so may look slightly different to other spreadsheet software. This is most likely to be the case with the conditional formatting part of the task so if using different software you might choose to demonstrate this. • Students should identify that DJ B plays the song least frequently. Linking the tables and adding the colour scale will help to highlight this. • You could ask students about the colours they selected and the significance of them. Did they select colour combinations that looked appealing or were they based on associations with high/low values?

Activity	Instruction	Guidance
4	• Ask students to complete (📖✏ Page 82) Task B, 'Highlight scores'.	• The Workbook activity takes students through a single colour conditional format. On this occasion, they apply the formatting to the entire row. • Encourage students to spend time experimenting with the formatting settings to customise their results. • The conditional formatting rule needs to highlight the full table 'A2:B21'. • The rule should always reference column 'B'; therefore it needs to be written '$B'. • The formula for the rule should be written as follows "=$B2 > 70%". The leading '=' is needed.
5	• Reflection: In pairs or small groups, ask students to discuss 'Is highlighting the entire row useful or not? Why do you think this?' and then complete (📖✏ Page 82) 'Reflection'.	• When thinking about whether full row highlighting is useful, there are a number of ways to look at it. It can be very useful when needing to refer to other columns to get a full picture of the data. If, in this example, the question was 'Which students scored over 70%?' then it would be useful to have the students highlighted. • Occasions when full row highlighting is less useful are when the table would become a mass of multiple colours that could be overwhelming or distracting. Also when you want to add different conditional formatting rules to different columns, which would you choose to highlight?
	Homework (Optional) (📖✏ Page 82) • Students are asked to create a spreadsheet of a collection of items they have at home. They need to include a primary key.	Examples: • books with the ISBN as the primary key. It would be even better if they added a number to represent their specific copy • websites with the domain name as the primary key • toys with a unique number assigned to each toy.

Activity	Instruction	Guidance
Build on this: Students could take the 'Station 1' spreadsheet tab and apply other conditional formatting rules to it. What happens when they apply a colour scale conditional format rule *and* a single colour entire row rule to the same table? What other formatting can they apply within the conditional formatting settings? Can they write conditional formatting criteria for each column that would be useful when answering data-driven questions? Are there accessibility issues to think about when using conditional formatting?		

Chapter 7:	Lesson 3 Search and capture data		
Student's Book page 115			
Learning Objective/s	• Describe how to search databases using single criterion, such as <,> • Evaluate the effectiveness of using data capture forms		
New keywords	**Database query** – Search tool that returns records that match the specified criteria **Data capture form** – Form designed to collect data from people, often digitally		
You will need	• A pre-prepared form (Google Forms is a good option for this). The form should have a couple of questions and you should have 'Link to Sheets' selected.	**Key pedagogy**	Develop practical skills

Activity	Instruction	Guidance
1	• Show an example of searching by filtering results in a spreadsheet. Explain that searching is also possible in a database by defining a database query, and use the examples (Page 115) to show how returned results are restricted. • Ask students to complete (Page 83) Task A, 'Searching databases'.	• You can run a class discussion to talk about the answers to the Workbook activity. • Students might include birds where the value is equal to not less/greater than the query. • The last question has no results. Explain it is possible to run a query and have no eligible matches returned.

Activity	Instruction	Guidance

2	• Talk through the uses of data capture forms (Pages 115 and 116) and ask 'Have you ever filled in a data capture form? If so, when?'
	• Talk through the tips for creating effective forms and highlight the Stay safe! advice.
	• Ask students to complete (Page 83) Task B, 'Effective forms'.
		• Students have used data capture forms in their computing lessons in earlier stages. They may also have used them in surveys at school or when signing up for things such as a library or event/club.
		• When talking through the tips, ask students why these things are useful.
		  ◦ What would they feel like filling in a form that was too long or where they didn't know how far through the form they were?
		  ◦ How easy would it be to analyse free-text data compared with checkboxes or radio buttons?
		  ◦ What if people made spelling mistakes so the data wasn't returned in database queries?
		• Why is it useful to have expected answer tips? Ask students to write the date on a piece of paper. Did everyone write it in the same format?
		• Expected answer tips are sometimes called 'data input masks'. These masks are written examples inside the input area but are usually a lighter colour and can be typed over.
		• Why is a confirmation message useful? Answer: It lets users know their submission was successful and it's polite. It's also a chance to add another message for users to read, for example, to advertise an event or product.
		• Students should identify that form 1 is the best form. They should have circled: the progress bar, expected answer text, checkboxes and 'skip' text.

Activity	Instruction	Guidance
3	• Demonstrate your online form to students. Load, complete and submit the form. • Show students your editing view of the form and click on the 'Responses' tab. Talk students through the tab and draw their attention to how your responses have been visually presented. • Click on 'View in Sheets' to show students your response in the linked spreadsheet. • Get students on to computers and share the link to your online form. • Ask students to complete the form and submit their answers. • Keep the 'Responses' tab and spreadsheet open on your computer so that students can see the response update in real time.	• Your form can be about any topic you feel is relevant. Ideally your form should include a free text, radio button and checkbox answer. • You could create an example similar to the 'School garden volunteering' example from the Student's Book and add additional questions such as: the types of club wanted, the school years that should be able to attend, and how long the sessions should be. • Students may query the difference between checkboxes and radio buttons. Use this as an opportunity to tell them that radio buttons are used when users can give only one answer. • You will be able to easily tell how many responses have been submitted so you can use this to encourage all students to submit their answers.
4	• Give students a link to your Forms app and ask them to complete the Build activity (Page 117). • Reflection: In pairs or small groups, ask students to discuss 'Which tips did you use when creating your data capture form?' and then complete (Page 84) 'Reflection'.	• The Build activity has screenshots for Google Forms, but if you are using a different app the students may need you to demonstrate the process. • Encourage users to experiment with the settings to customise the appearance of their forms and to make the best user experience. • Make sure that students are testing their forms. They can do this multiple times with different responses to test them fully, and they should test them after each big change. • Students are performing User Experience (UX) tasks here. UX is a key role in industry – it is important that organisations gain user input when they are designing user-facing experiences such as data capture forms.

Activity	Instruction	Guidance
	Homework (Optional) (Page 84) • Students need to ask a trusted adult about a time when they had to fill in a data capture form.	• You could provide an example of your own experience of using a data capture form to support students who don't have a trusted adult available to ask. • This activity helps students to develop an understanding of why it is important to follow best practice when designing data capture forms.

Build on this: Get students to test their forms with their peers. They should watch how their peers interact with their forms and ask them about their experiences, so that they can improve the usability of their forms.

Chapter 7: Lesson 4 Ask your audience

Student's Book page 118

Learning Objective/s	• Design and create a data capture form • Capture and store data		
You will need	• Access to your chosen online forms app to create a new form	**Key pedagogy**	Develop practical skills

Activity	Instruction	Guidance
1	• Introduce the project brief and outline for the project (Page 118). • Talk through the questions students should include when designing their forms. • Ask students to complete (Page 85) Task A, 'Design your form'.	• Students can work in groups and discuss what they would like to spend the money on, to help develop their questions. As you observe them designing their forms, make sure that they are taking into account that not all people will have the same interests as they do. • You can give students a specific budget to spend in your local currency to help them to identify the types of things they can afford to buy. They may decide to split the money across different areas or buy one big thing. • The tips for effective forms from last lesson are repeated in the Workbook for reference. Students could also look back to their Workbook tasks from the last lesson to see examples of good design.

Activity	Instruction	Guidance
2	• Give students the link to the online form app, and ask them to create their forms. • Encourage students to test their forms to make sure they work as expected.	• Encourage students to use their own forms to test that they work as expected • Students should check their test responses in both the responses tab and the linked spreadsheet.
3	• Ask students to start capturing data from fellow students. • Tell students that they will have until next lesson to capture data. • Reflection: In pairs or small groups, ask students to discuss 'Did you have any challenges designing your form? If so, describe them' and then complete (Page 85) 'Reflection'.	• Students should gather data from all other students in their class. • You could encourage students to also gather data from other classes by speaking to teachers and asking if they can briefly interrupt their class so their students can complete the form. Organise this so that classes are only interrupted once each.
	Homework (Optional) (Page 85) • Students are asked to describe how they would improve their data capture forms if they had more time.	• Other students may be confused when answering some questions. This could be addressed by rewording questions or adding tips or descriptions. • They may have used check boxes when they should have used radio buttons.
	Build on this: Students could create a poster for their project with a link to the form and then place it in a communal area within the school, or share it online using the school's virtual learning environment (VLE). Students from the school can then complete the form at any point during the next week.	

Chapter 7:	Lesson 5 Analyse your data		
Student's Book page 119			
Learning Objective/s	• Use conditional formatting to help to analyse data and make decisions		
You will need	• The spreadsheets linked to the data capture forms	**Key pedagogy**	Develop practical skills

Activity	Instruction	Guidance

1	• Talk through the conditional formatting examples (Page 119).
	• Ask students to complete (Page 86) Task A, 'Create your model'.
		• Highlight the different ways that conditional formatting has been used in the spreadsheets.
		  ◦ In the first, it is used to make the most frequently occurring choice stand out.
		  ◦ In the second, it has been used to work out the average amount people say should be spent on one area, to help split the money fairly.
		  ◦ In the third example, multiple rules have been created to show how age impacts the suggestion.
		• Remind students that they will be presenting their spreadsheets in the next lesson so the conditional formatting should help them make their decisions but also help others to see how they reached that decision by making the relevant cells stand out.
		• Students may choose to use their knowledge of formulae to add extra columns and rows to their data. They could also use their knowledge of standard formatting to make their spreadsheet look more appealing and accessible.

2	• Ask students to complete (Pages 86 and 87) Task B, 'Student-led decisions'. • Students should spend any remaining time preparing their presentations for the next lesson. They can practise what they are going to say and how they are going to showcase their projects.	• Students should answer the questions in the Workbook activity. Their answers will be useful next lesson when they present their projects, so they should be answered thoroughly. • Encourage students to practise their showcase aloud. Each group should consider whether they will be sharing their form and spreadsheet. They should also consider how they will talk through their decision and explain why it is the right one based on the data they have collected.

Activity	Instruction	Guidance
3	• Reflection: In pairs or small groups, ask students to 'Share the challenges that you had while creating your model' and then complete (Page 87) 'Reflection'.	
	Homework (Optional) (Page 87) • Students create a presentation with 1–3 slides to support the showcase in the next lesson.	• Remind students how to create a presentation using the software that is available to them.

Build on this: Students could create a slideshow with screenshots and bullet points to showcase their form and their spreadsheet. They should annotate the slides to show the key functionality they used and justify why it helped them to reach a decision.

Chapter 7: Lesson 6 Present your decision

Student's Book page 120

Learning Objective/s	• Present your analysis and decision • Evaluate the effectiveness of your data capture		
You will need	• Students might need access to their forms, spreadsheets and any other resources created in the last few lessons that will help them to showcase their decisions	**Key pedagogy**	Develop practical skills

Activity	Instruction	Guidance
1	• Explain that this is a showcase lesson in which students will present their findings and the process they used to reach their decisions (Page 120).	• Conduct a class showcase. • Ask students to talk about the questions they asked in the form, and the functionality they used to make sure that their forms were effective. • Encourage students to talk about who they asked to complete the form and how they went about it. Did they also include testing of the form? • Students should be able to answer questions about their conditional formatting scheme and why they used it. They should talk about why it helped them to arrive at a decision and why it was useful. • Students should be able to summarise their decisions. Did all students reach the same decision or were there differences?

Activity	Instruction	Guidance
2	• Ask students to complete (Page 88) Task A, 'Reflection'. • Ask 'What improvements did you identify that you would like to make to either your form or your spreadsheet?'	• Students should reflect on the full process including whether or not they are confident they met the project brief. • During the discussion identify trends in the answers and discuss these further.
3	• Conclude with the celebration and a recap of everything they have achieved this lesson. • Ask the reflection question: 'Why are spreadsheets useful tools for making data-driven decisions?' • Ask students to record their reflections in (Page 88) 'Reflection'. • Students can then complete the (Page 88) Project skills checklist.	• Spreadsheets are useful because the data is organised and structured in a way that makes analysis possible. • Data can be entered manually or automatically and then manipulated using calculations, formulae and formatting.
	Homework (Ooptional) (Page 88) • Students are asked to summarise the experience they have gained during this chapter. How will it make them think differently next time they fill in or design a data capture form?	• Students should consider what features make a well-designed data capture form for collecting data that can then be analysed. What makes a data capture form easy to use? How should the questions be worded? When is it appropriate to use checkboxes or radio buttons? Is help text provided?

Build on this: Survey the teachers and school staff using the students' data capture forms. Compare the data obtained by students and staff. Does analysis lead to the same or a different decision? What similarities/differences are there?

Chapter 8 – Create with code 3

Project: Plan, design, test and build a device to welcome visitors to your classroom

Chapter overview

In this chapter, students will explore the importance of planning in project development. They will look at user-centred design and evaluate prototypes based on user needs. They will put their knowledge into practice by designing a device that makes their classroom a welcoming space. Finally, they will showcase their devices alongside the testing process.

Chapter 8: Lesson summary		
Lesson	**Learning Objectives**	**Summary**
1	• Define project plans and their use in software development projects • Compare design of a range of different devices	Students are introduced to the concept of having a project plan and its uses in the software development field. They explore common traits that lead to project failure and learn how to avoid them. Finally, they learn about user-centred design and how to consider the user when designing projects.
2	• Describe how to identify errors in a program by using a test plan	Students discover the importance of creating a formal test plan to make sure that their projects do not contain errors and that they meet the user requirements. They are given two different micro:bit projects to test using a test plan.
3	• Evaluate prototypes for software and/or hardware development projects	Students explore how prototypes are used in software development and assess the quality of example prototypes. Then they explore the chapter project and develop their own prototype sketches for their 'classroom welcome' device.
4	• Develop a program for a physical computing device to generate multiple outputs, based on multiple inputs	Students are given micro:bit code snippets and examples to help them to create the code for their 'classroom welcome' devices.
5	• Apply test plans • Systematically identify and debug errors in text-based programs	Students create a test plan for their code to make sure that their devices work as expected. They then test and debug their code.
6	• Showcase a device and test plan	Students prepare and perform a showcase of their classroom welcome devices. Then they reflect on their experiences of creating the devices and share their reflections with a partner.

End of chapter project	**Plan, design, test and build a device to welcome visitors to your classroom**
Example ideas that students could design:	• A micro:bit on the wall next to a classroom door. When someone presses button A it produces musical notes. • A micro:bit on a table next to a classroom door. When someone presses a button, an image is displayed on the LEDs. • A micro:bit hanging from a string near a classroom door. When someone swings the micro:bit, an image is displayed on the LEDs.

Why we are learning this	
Useful skills covered in this chapter	• Following a project from planning to showcasing gives insight into how the apps and devices that they use are developed
Links to future careers	• Project manager, tester and software developer are different job roles in the tech industry, which appeal to different people
Links to other subjects	• Maths – calculation devices • Business – planning a project

Chapter 8 – Create with code 3

Project: Plan, design, test, and build a device to welcome visitors to your classroom

Name	

	Working towards (1)	Meets expected (2)	Exceeding expected (3)	Score
Welcome device	• A micro:bit welcome device with input(s) or output(s)	• A micro:bit welcome device with inputs and outputs	• A micro:bit welcome device with a range of different inputs and outputs	
Prototype	• A prototype has been sketched	• A prototype has been sketched that meets the brief	• A prototype has been sketched that meets the brief and has been improved based on feedback	
Test plan	• A test plan has been partially created but it does not test all functionality required for the brief	• A test plan has been created and testing confirms the device meets the brief. Observations and feedback have been documented	• A test plan has been created, testing confirms the device meets the brief. Useful observations and feedback have been elicited and documented	

Teacher feedback	
Learner response	

Chapter 8: Lesson 1 Project plans

Student's Book page 123

Learning Objective/s	• Define project plans and their use in software development projects • Compare design of a range of different devices		
New keywords	**Project plan** – Carefully planned tasks that ensure a project meets a deadline to the correct requirements **Software development** – Creating software (programs) for a specific user requirement or need **Scope creep** – Features that were not in the original requirements getting added over time **User-centred design** – Involving the end user of a product in every stage of the design process		
You will need	• A BBC micro:bit device to show the class (optional) • The BBC micro:bit feature overview <u>(Website C8L1R1)</u> (optional)	**Key pedagogy**	Classroom talk

Activity	Instruction	Guidance
1	• Introduce the project for this chapter to the students (Page 122). They will be creating a device that will welcome visitors to their school classroom. They will be using a micro:bit to create their device. • If available, show a BBC micro:bit device to your class to prompt recall. You could instead show an image of the BBC micro:bit version you will be using from the feature overview link (Page 123). • Ask 'What projects have you used a BBC micro:bit for? What inputs and outputs did you use?'	• You could remind them of other micro:bit projects that they have completed before. • Use the responses to fill in gaps in the students' knowledge of the micro:bit.
2	• Introduce the concept of a project plan and why it is important to have a plan before starting any large project. Use the cartoon strip (Page 123) to demonstrate that this character did not have a plan and ended up with a chair that would not work effectively for the user. • Ask 'Can you think of a time when you have tried to complete a project without a plan? Was it still successful? Would a plan have helped you?'	• A prompt for this might be that they didn't plan enough time to complete a piece of homework so it was rushed and got a low grade.

<table>
<tr><th>Activity</th><th>Instruction</th><th>Guidance</th></tr>
<tr><td>3</td><td><ul><li>Explain how project plans relate specifically to software development (Page 124). Project plans are used in real-world software development. Talk through the list of typical features that may appear on a project plan.</li><li>Ask students to complete (Page 89) Task A, ‘Order the tasks’.</li><li>Discuss responses to the ordering task and provide the correct ordering.</li></ul></td><td><ul><li>Some details of the ordering may depend on the specific project. See the answers for more details.</li><li>Note that there is often overlap between steps in a real project and steps may be repeated for new features.</li><li>Relate students’ answers to projects they may have completed in Computing or other subjects such as Business.</li></ul></td></tr>
<tr><td>4</td><td><ul><li>Read through ‘The new computer game’ case study (Page 125).</li><li>Ask ‘There were two main factors that made this project fail. What were they?’</li><li>Discuss the list of the main reasons why projects fail (Page 125). Make sure that students understand the terms used and ask them if they can think of some examples of each.</li></ul></td><td><ul><li>The main reasons the new computer game failed were:<ul><li>Not enough time was given for the project to be successful.</li><li>The game needed to be tested before going to the customers. The game needed to be ready for testing on time in order for this task to be completed successfully.</li></ul></li></ul></td></tr>
<tr><td>5</td><td><ul><li>Ask students to complete (Page 90) Task B, ‘Case study: Mobile ticketing’</li></ul></td><td><ul><li>The main reasons the mobile ticketing failed were:<ul><li>missing stakeholders</li><li>scope creep.</li></ul></li><li>This is a good chance to mention digital exclusion. Failure to talk to the customer stakeholders led to a group without smartphones from being digitally excluded. You could ask ‘What could they have done to prevent this whilst still reducing plastic?’ Answer: They could have a plastic solution only for those customers that needed that option.</li></ul></td></tr>
</table>

Activity	Instruction	Guidance
6	• Introduce students to the concept of 'user-centric design' (Page 126). This is where the user is involved throughout all stages of the development process. The example uses a fast-food restaurant ordering system which they may have seen in the real world. • Explain that by interviewing, observing and surveying customers, the company realised that a translation feature would be beneficial for customers.	• You could ask students here if they have ever seen a screen like this in a restaurant and if they had any problems using it. For example, was it faster? • Make links to previous chapters where students designed a feedback kiosk and designed a data capture form.

7	• Ask students to complete (Pages 90 and 91) Task C, 'Effective design'.	• Option b, in the middle, is the design that would best suit their customer needs. The reasons for this are:
		  • The writing and the buttons are big and clear.
		  • There are no unnecessary links cluttering the page.
		  • The position of the search bar and buttons are close together so can be selected with one hand.
		  • There is the option to type or speak.

8	• Reflection: In pairs or small groups, ask students to discuss 'Share the key things you learnt from this lesson. What did you find most interesting?' and then complete (Page 91) 'Reflection'.	• They could share answers with a partner or with the whole class.
	Homework (Optional) (Page 91) • Students are asked to interview a friend or family member about an app or device that they use.	• You could pair students up to interview each other or suggest a trusted adult in school. • Key learning: Apps and devices that we use could be improved if more time was spent on understanding user requirements.

Build on this: **Invent a new game**

What would be your key requirements for the new game? How would you understand the needs of your end users?

<table>
<tr><th>Chapter 8</th><th colspan="3">Lesson 2 Test plans</th></tr>
<tr><td colspan="4">Student's Book page 127</td></tr>
<tr><td>Learning Objective/s</td><td colspan="3">• Describe how to identify errors in a program by using a test plan</td></tr>
<tr><td>New keywords</td><td colspan="3">Iterative testing – Running your program as you develop it to check that there are no syntax (or block) errors
Test plan – List of tests to be carried out on a program to ensure that it works as expected</td></tr>
<tr><td>You will need</td><td>• Access to the EduBlocks platform for individual or paired work
• Access to the EduBlocks/micro:bit projects
  • Addition: <u>(Website C8L2R1)</u>
  • Multiplication: <u>(Website C8L2R2)</u>
  • Printed copies of the 'Using a micro:bit and the micro:bit simulator with EduBlocks' worksheet (optional)</td><td>Key pedagogy</td><td>Develop practical skills</td></tr>
</table>

Activity	Instruction	Guidance
1	• Start the lesson by reminding students that they have identified and fixed errors in the programs that they have created before. This process is called debugging. • Introduce the concept of a test plan (Page 127) to the students. This is a more structured way of testing whether a program works. It should list the tests required to make sure that the program works as expected. • Talk through the example test (Page 127). Test plans should list the test number, provide a description of the test, state the input that the user might enter and state the expected result. • Explain that 'Pass' means that the test passed and no fix is needed. • When the test plan is being used, the tester should enter the input and see if the expected output matches the result. If there is an error, then this should be identified and fixed.	• The test plan used in the example is not complete, it is just a short example of part of a test plan. Students may ask if there should be a fourth test that checks whether the answer is correct. This is absolutely true and is an excellent observation to make. They will see a complete test plan in the next activity.

Activity	Instruction	Guidance

2	• Instruct the students to complete (Page 92) Task A, 'Follow a test plan'.
	• Explain that students will be testing projects for the micro:bit that have been created using EduBlocks.
	• Open the Addition project on your computer and demonstrate how to:
	  ◦ Click 'Simulator' to enter the Simulator
	  ◦ Click the Play arrow to run the project
	  ◦ Press the A and B buttons
	  ◦ Choose 'shake' from the Accelerometer drop-down menu and click 'Send'.
	• Students will need access to the Addition project in EduBlocks. They should follow each test carefully, identify any errors and fix them.
		• This is a good activity to complete in pairs because they can support each other in identifying and fixing the errors.
		• You could ask students to check the answers using their calculator or the calculator built into a search engine.
		• On a lower resolution screen, the Accelerometer menu may not be visible, zoom out (Ctrl-+) until it is visible.
		• Students used EduBlocks to write Python code in Chapter 5, Create with code 2. Students may notice that the micro:bit code in EduBlocks is also Python code.
		• You can also give students a copy of worksheet 'Using a micro:bit and the micro:bit simulator with EduBlocks' to help them run code of the EduBlocks simulator.

3	• Once students have completed the test plan, bring the class back together to discuss where the errors were and how they fixed them.	• Use the example in the answer sheet to support this discussion.
4	• Now instruct students to complete (Page 93) Task B, 'Write and follow a test plan'. This time, students have been given a program that they will create a test plan for. They will need access to the Multiplication project in EduBlocks. Once they have created the test plan, they should perform the tests and identify and fix any errors. • Instruct students to read the requirements in the bulleted list carefully so that they are sure of what the program is expected to do. They should use this list to help them to write their test plans.	• An example completed test plan has been provided on the answer sheet. You can use this to help you support the students if needed. • This is another activity that may benefit from paired work.
5	• Once students have finished their testing, ask them which errors they found and how they fixed them. Also ask if they missed any tests out and had to add them in during their testing phase.	• This is a good time to address any pitfalls that students might have gone through during this activity. They may have not been specific enough with their tests or not broken them down into enough steps.

Activity	Instruction	Guidance
6	• Reflection: In pairs or small groups, ask students to discuss 'Why is testing important?' and then complete (📖✏ Page 93) 'Reflection'.	• You could link back to Lesson 1 where they discovered that lack of testing can cause user frustrations and ultimately lead to losing customers and money. • Students will need the micro:bit EduBlocks worksheet for future lessons. Either collect them in or ask students to store them in their books.
	Homework (Optional) (📖✏ Page 93) • Students research a famous software error from a space mission.	• You could direct students to a specific error that is relevant to them.

Build on this: **Improve the micro:bit projects**

Students could pick one of the games from the micro:bit projects used in this lesson to improve. They could add a new welcome message, add a new end message, change the range of numbers generated, provide feedback, or add more questions. They should update the test plan by replacing a test or adding a new test.

Chapter 8: Lesson 3 Prototypes

Student's Book page 129			
Learning Objective/s	• Evaluate prototypes for software and/or hardware development projects		
New keywords	**Prototype** – Sketch or model to test a concept for a project		
You will need	• Coloured pencils for the students to add colour to their prototype sketches (optional)	**Key pedagogy**	Get creative

Activity	Instruction	Guidance
1	• Introduce the concept of prototyping to the students (Page 129). • Show the students the example prototype sketches for the music app. • Ask 'Do you think that the sketches show the key requirements for the project? What could be improved?'	• Give students time to read the requirements carefully and study the sketches. • They should notice that it is not easy to tell what the icon used on the home screen does. This type of icon is usually used to shuffle or repeat songs and users may not be sure what its purpose is here. • The other key thing to notice is the play button on the second screen. The requirements state that the songs should play automatically, so it would be more useful to have a stop button here instead of a play button.
2	• A further prototype sketch has been provided in the Workbook to give students another chance to comment on whether it meets the requirements. • Ask students to complete (Page 94) Task A, 'Does it meet the requirements?' • They should carefully read the requirements and review the images and annotations. They should decide whether or not the design meets the requirements and, if not, state what is missing.	• Model answers have been provided in the answer section. You can use this to guide your support for students if necessary.
3	• Reintroduce the chapter project to your students using the brief (Page 130). • Ask students to think about what they might want to do for their project. • Instruct students to complete (Page 95) Task B, 'Your prototype sketch' to create an initial prototype sketch of their 'classroom welcome' devices. • After they have had time to do this, ask students to sit with a partner to look at each other's designs. They should check whether their sketches show that they have met the requirements and they should give feedback on how they might improve their designs. • After they have each received feedback, they should complete (Page 96) Task C, 'Your improved prototype sketch' by doing another prototype sketch based on their feedback.	• You could also provide feedback if extra support is needed in this area. • If there is time, students could add colour to their final prototype sketches.

Activity	Instruction	Guidance
4	• Reflection: In pairs or small groups, ask students to discuss 'Why is prototyping important when creating a new product?' and then complete (Page 96) 'Reflection'.	• Prototyping is useful because it allows you to check that you are meeting the requirements. It can also save time and money because you can create quick versions for user testing before spending time on a final product.
	Homework (Optional) (Page 96) • Students are asked to describe how a change in requirements would affect the design for their devices.	• Key learning: Different requirements, such as a change in the user needs or the environment in which the project will be used, can result in significant changes to the design.
Build on this: **Add colour or start thinking about code** Students could either add some colour to their sketches or they could start to plan out the code that they might need for their classroom welcome device.		

Chapter 8: Lesson 4 Create your device

Student's Book page 131

Learning Objective/s	• Develop a program for a physical computing device to generate multiple outputs, based on multiple inputs		
You will need	• A screen that can access the EduBlocks links for showcasing the example student projects • Printed copies of the 'Creating a micro:bit project with inputs and outputs with EduBlocks' worksheet • Music example (Website C8L4R1) • Face up example (Website C8L4R2) • String example (Website C8L4R3)	**Key pedagogy**	Get creative

Activity	Instruction	Guidance

1	• Remind students of the project brief.
	• Share the examples of students' projects (Page 131).
	• Ensure students have access to copies of the micro:bit EduBlocks worksheet.
		• As you talk through the examples, you can demonstrate how each device works using the micro:bit simulator for these projects.
		  • Music example (Website C8L4R4)
		  • Face up example (Website C8L4R5)
		  • String example (Website C8L4R6)
		• Explain that you are reading the code to look for the inputs that you need to provide to see the different outputs.
		• When demonstrating the accelerometer features, such as 'shake' and 'face up', you will need to update the 'Accelerometer' drop-down under the simulator, then click 'Send'.
		• Students may notice that the code is all in one loop instead of separate 'on event' scripts as they used in the MakeCode editor. This is because EduBlocks matches the Python code you would need to write.

2	• Introduce the Build task and split the students into their groups from last lesson. • Advise students that they will need to refer back to their prototypes from last lesson. • Students will need to use the computers and the EduBlocks website. • Students should complete (Page 97) Task A, 'Plan your build' as they explore EduBlocks and create their code.	• Students can work on the micro:bit simulator until their code has been finalised. • Students should save their code as they will need it in the next lesson. • Remind students to import their libraries as it's easy to forget this when thinking about the code for the inputs and outputs.
3	• Reflection: In pairs or small groups, ask students to discuss 'How successful were you at sticking to your prototype without scope creep?' and then complete (Page 97) 'Reflection'.	• Did all students stick to the prototype they designed last lesson or when seeing the examples and experimenting in EduBlocks did they decide to add more functionality? • Did any additional features mean that they didn't finish in time or were these managed within the time allowed for the build?

Activity	Instruction	Guidance
	Homework (Optional) (📖 Page 97) • Students imagine that a friend has suggested that it would be brilliant to add lots of additional features to their design. They need to consider whether or not this is a good idea at this stage of the project.	• Students should recognise that they need to avoid scope creep to ensure that they can deliver the project on time.

Build on this: Students could share their design pages from their Workbooks with other groups. What were the key parts that they noted or sketched that helped when building? Did they use it mainly for reminders of things to include in the code or was it more of a visual map of what each input led to? If they had one 'top tip' what would it be?

Chapter 8: Lesson 5 Test and refine your device

Student's Book page 132

Learning Objective/s	• Apply test plans • Systematically identify and debug errors in text-based programs		
You will need	• Students will need access to micro:bits. or the micro:bit simulator in EduBlocks • Students will need access to their saved projects from the last lesson • Printed copies of the 'Creating a micro:bit project with inputs and outputs with EduBlocks' worksheet	**Key pedagogy**	Develop practical skills

Activity	Instruction	Guidance
1	• Explain that today's lesson focuses on testing. Students will be testing their own devices first. • They need to test that the device works as planned and that it meets the brief. • Ask students to complete (📖 Page 98) Task A, 'My test plan'.	• At the beginning of this lesson, students should have a finished project to test. • The first round of testing can be completed with the simulator. • Students can look back to the test plans in their Workbook from Lesson 2 for a reminder of what types of tests to include.
2	• Students should make any necessary changes to their devices based on the test results. • Demonstrate how to download code onto a physical micro:bit. • Give students time to test their code again on the simulator before downloading to a physical micro:bit.	• Encourage students to run through their test plan again once the code is on the micro:bit. • All students in a group should have chance to experience using the physical device.

Activity	Instruction	Guidance
3	• Students should now take part in an observation and feedback session. • Ask students to complete (📖✏ Pages 98 and 99) Task B, 'User research'.	• Pair groups together for user research sessions. • When it is their time to use another device, encourage each member of the group to try the other group's device and provide feedback. • When it is their time to observe another group using their device, they should all make observations and take feedback notes. • Give groups time at the end to discuss their notes within their own groups. Were there common themes? Did anyone spot something valuable that no-one else did?
4	• Students should make any necessary changes to their devices based on their observations and feedback from the other group.	• Make sure that students are taking feedback on board. They do not have to make all changes if they feel strongly that the feedback is not representative of most students, but if they are not going to act on it, then they should be able to explain why.
5	• Reflection: In pairs or small groups, ask students to discuss 'What differences did you notice between testing your device in the simulator and testing your physical device with end users?' and then complete (📖✏ Page 99) 'Reflection'.	• Sometimes with physical devices that use inputs it is a different experience when you actually use a physical device rather than press an onscreen button. • There might be accessibility issues with the inputs, or the outputs might not look and sound as you expected. • Perhaps tweaks to brightness, volume or pitch needed to be made, or maybe the position in the room created issues that could not have been found with the simulator.
	Homework (Optional) (📖✏ Page 99) • Students write instructions for using their devices.	• A device should be easy to use without complex instructions. The instructions should be short and easy to understand.

Activity	Instruction	Guidance
Build on this: Invite users from another year group, or staff from within the school, to enter the classroom and test students' devices. Do students make the same observations and receive the same feedback from people of different age groups or those unfamiliar with the project and micro:bit? They should make sure they address any feedback for the audience they are making the device for.		

Chapter 8:	Lesson 6 Showcase your device		
Student's Book page 133			
Learning Objective/s	• Showcase a device and test plan		
You will need	• The completed 'classroom welcome' devices	**Key pedagogy**	Showcase

Activity	Instruction	Guidance
1	• Students should have now completed their 'classroom welcome' devices and tested them against a test plan.	• If they need more time to complete their devices then you could do this now and reduce the time for showcasing their devices.
2	• Go through the showcasing tips and instructions (📖 Page 133). • Give your students some time to think about what they would like to say and to practise their talk.	• Depending on the time, you may wish to ask students to showcase to a partner, small groups or the whole class. You could also invite a guest and ask them to pick some devices to test.
3	• Showcase the 'classroom welcome' devices.	• Highlight examples of students following the best practices for showcasing from the Student's Book.
4	• Ask students to complete (📒 Page 100) Task A, 'Reflection'.	• Emphasise that planning, testing and understanding user requirements can avoid problems later.
5	• Students share their reflections with a partner or in small groups.	• Explain that software and hardware projects often have big teams and are developed over many months or years. Celebrate what they have achieved in such a short time.

Activity	Instruction	Guidance
6	• Reflection: In pairs or small groups, ask students to discuss 'How well did your device create a welcoming environment for visitors to your classroom?' and then complete (Page 100) 'Reflection'. • Students can then complete the (Page 100) Project skills checklist.	
	Homework (Optional) (Page 100) • Students are asked to write (but not send) a letter to the developers requesting a feature for an app or device that they use regularly.	• They will need to explain why their feature is important. • The product developers will be working to a budget and schedule and will want to avoid scope creep.

Build on this: **Record their showcase**

You could film the students as they complete their showcase. This could then be edited together for a celebration-style event that all students could watch. You could even invite a special guest. When students see themselves on camera they can think about their performance and how they might improve it in future.

Creating a micro:bit project with inputs and outputs with EduBlocks

Navigate to EduBlocks and create a new micro:bit project.

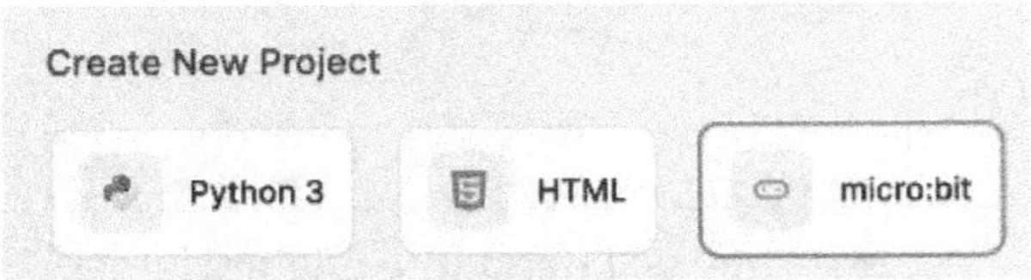

Add 'from microbit import *' from the 'Imports' section.

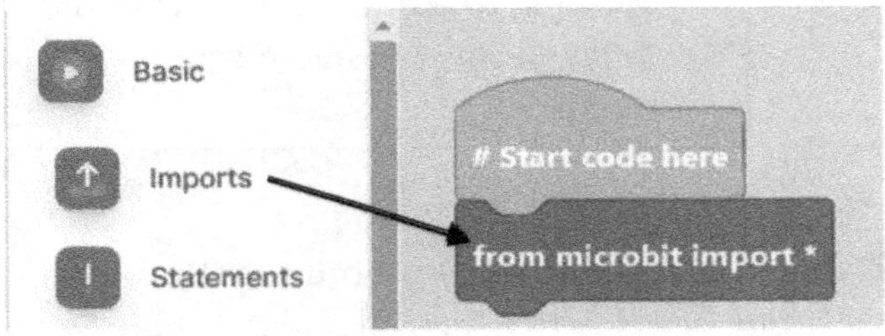

micro:bit inputs in EduBlocks

To run code when a button was pressed:

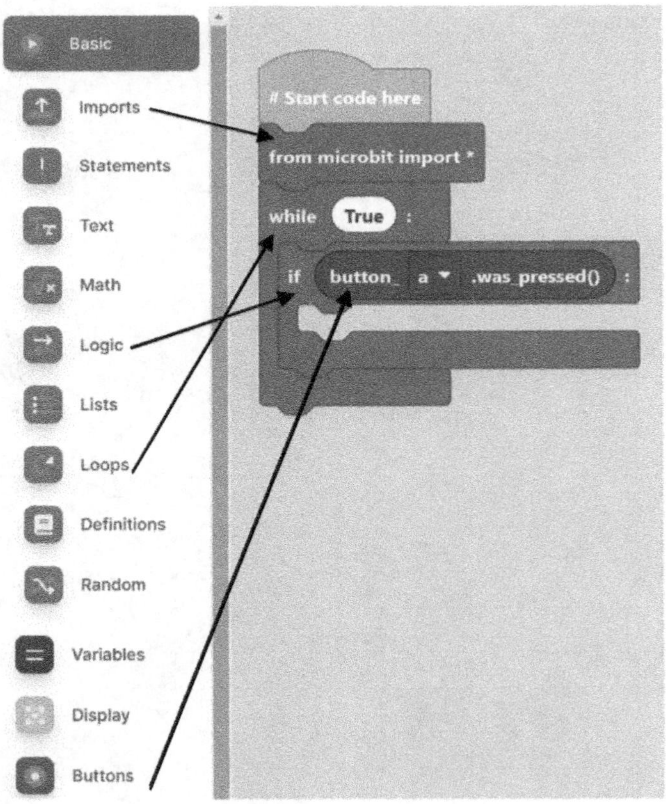

To run code when an <u>accelerometer gesture is detected:</u>

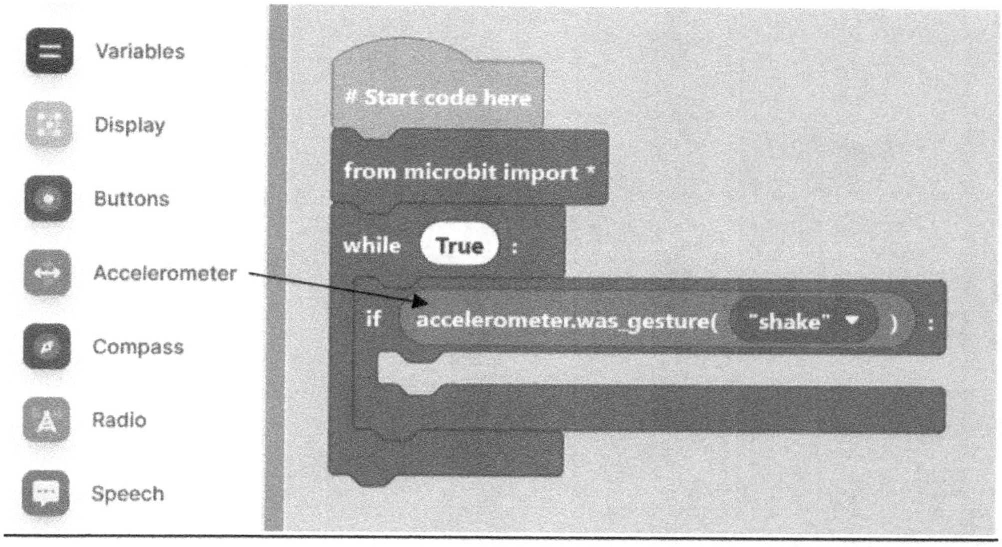

micro:bit outputs in EduBlocks

You can display numbers, text or images on the micro:bit LED screen using blocks from the 'Display' section:

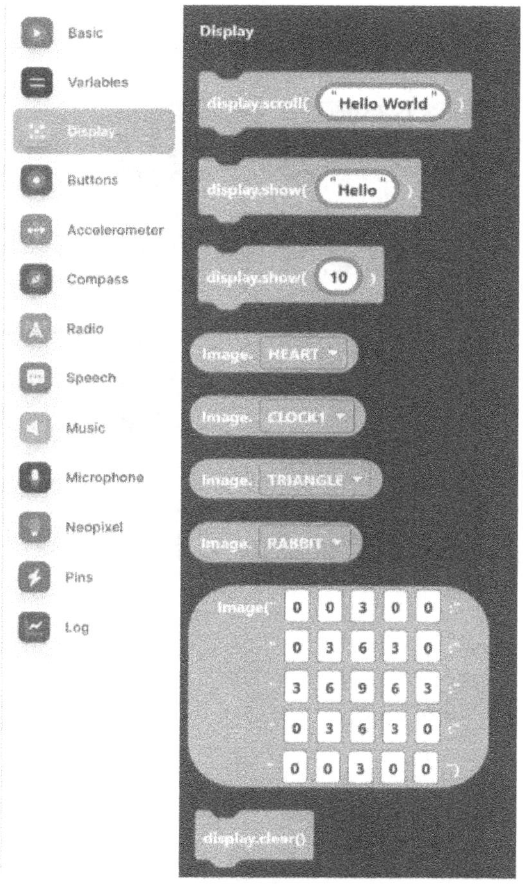

You can create an image with pixels with different brightness levels from 0 (off) to 9 (full brightness).

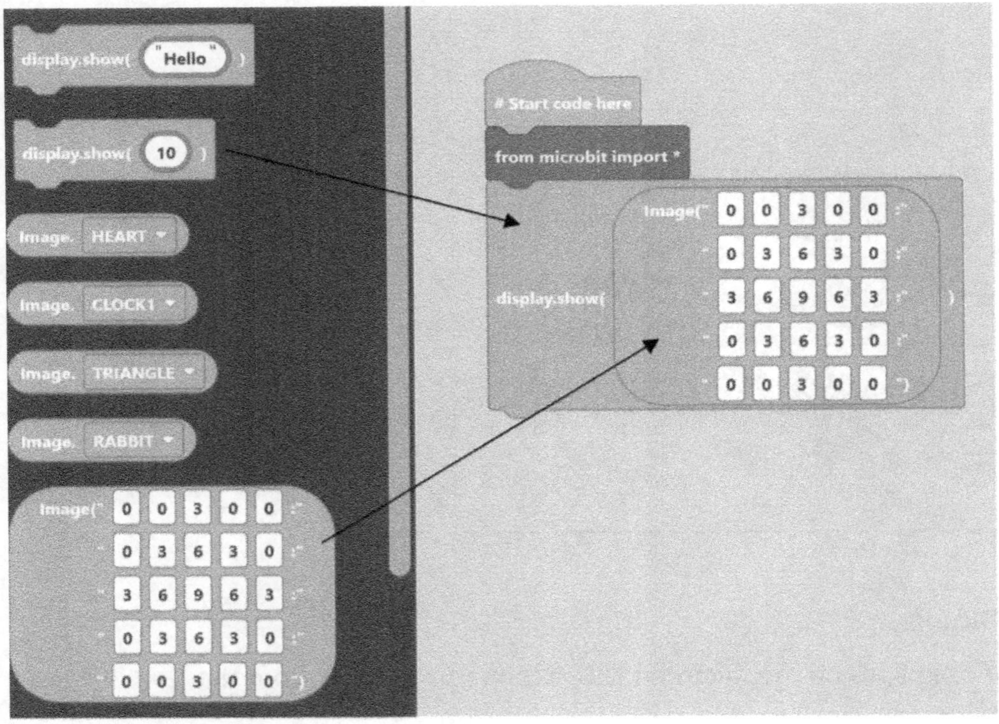

You can play music and sounds with the blocks from the 'Music' and 'Speech' sections:

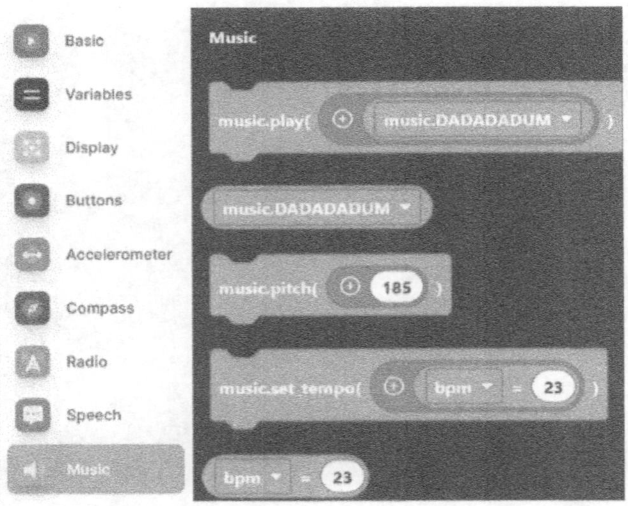

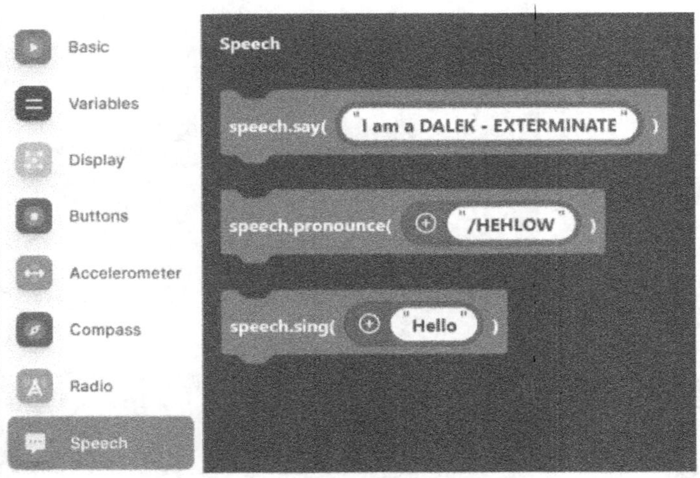

Using a micro:bit and the micro:bit simulator with EduBlocks

From your EduBlocks micro:bit project:

- Click on the 'Simulator' tab.
- Click the play icon to run your code.

Use the buttons under the micro:bit to stop or repeat your code.

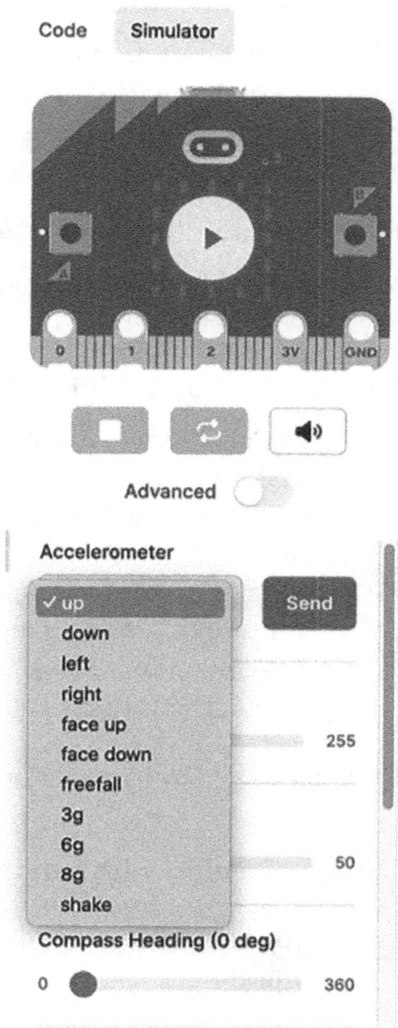

Inputs

If your code has inputs from buttons, you can use the buttons on the micro:bit.
Additional input options can be found under the micro:bit, such as the:

- 'Accelerometer' dropdown list
- micro:bit sensor sliders
- button and pin icons.

After selecting a value from the Accelerometer dropdown list, remember to click 'Send'.

Using a physical micro:bit

To transfer your code to a physical micro:bit either click on:

- 'Download' to download a file to your computer to transfer later

or

- 'Connect' then attach your micro:bit to your computer to directly transfer the file.

Answers to Workbook questions

Chapter 1 – Our Digital World

Chapter 1.1 Safe and healthy use of technology

Task A. Protecting your wellbeing online

Answers should show awareness of the need to protect mental and emotional wellbeing as well as physical wellbeing.

Task B. Stop and think!

1. Learners should make sure they know why access to their microphone is needed and make sure that they trust the site.
2. Learners should make sure they know why access to their location is needed and make sure that they trust the site.
3. It's a personal decision whether to 'Accept all cookies' on a website you trust. It may allow the website to provide a more useful experience. But you should be aware of privacy risks.
4. The option to 'Manage Cookies' gives you more control. If you have time then it is a good idea to look at the settings and get more understanding of how your data will be used.

Chapter 1.2 Responsible participation in online communities

Task A. Dealing with trolling behaviour

1. You **shouldn't** try and convince them to be nice
2. You **should** ignore them
3. You **shouldn't** be mean to them
4. You **should** report them
5. You **should** ask an adult for support if the behaviour upsets you

Chapter 1.4 Train a machine learning model

Task A. Complete the AI definitions

1. A machine learning **model** combines an algorithm and **training** data.
2. **Supervised** machine learning is when a human adds **labels** to data.
3. **Unsupervised** machine learning is when a computer detects **patterns** in data.

Task B. Labelling data

Examples could include: people, other cars and vehicles, traffic cones, animals, barriers, traffic signs, trees.

Chapter 2 – Content creation

Chapter 2.2 Usage rights

Task A. Image usage rights

1. Licence types

(a) Royalty Free – copyrighted image that can be used after paying a fee
(b) Creative Commons – the image can be used but the owner must be credited
(c) Copyright Free – can be used freely without the need to credit the owner

Chapter 3 – Create with code 1

Chapter 3.1 Logic gates

Task A. Boolean expressions

Circle the images that would result in a TRUE outcome from the Boolean expressions.

1. 'Black AND square'

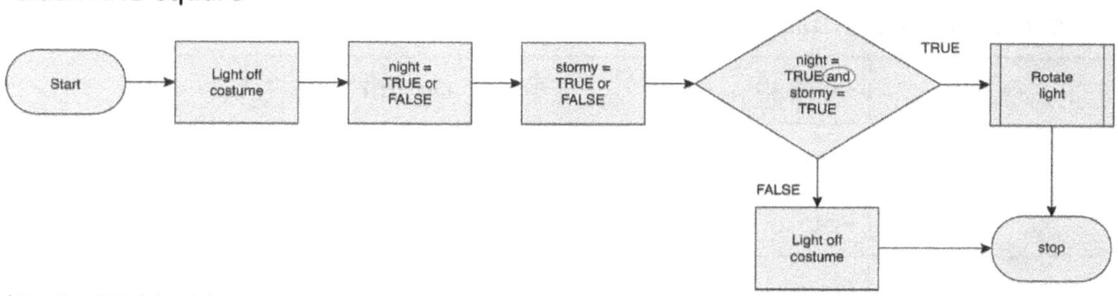

2. 'Circle OR black'

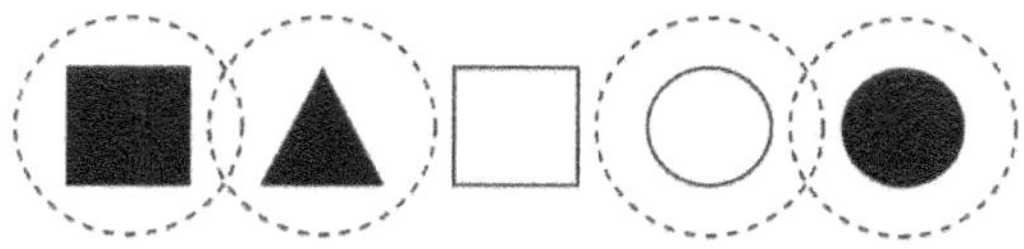

3. NOT 'Circle'

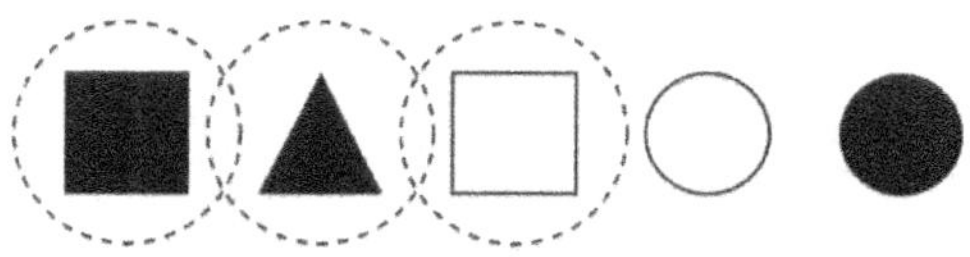

4. 'Less than four corners OR black'

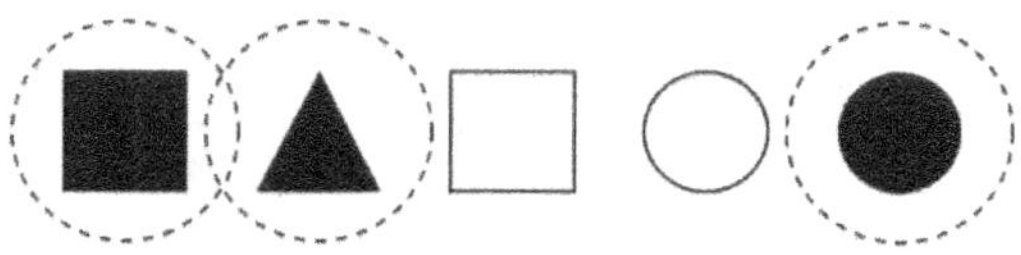

Task B. Investigate logic gates

1. ON
2. OFF
3. OFF
4. ON
5. ON
6. ON
7. OFF
8. ON

Chapter 3.2 Selection in flowcharts

Task A. Label the flowchart symbols

1. D
2. A
3. B
4. E
5. C

Task B. The correct ingredients

Ingredient 1	Ingredient 2	Output
Tomato	Onion	Soup bowl costume
Tomato	Potato	Soup bowl costume
Potato	Onion	Soup bowl costume
Carrot	Potato	Slime bowl costume
Carrot	Tomato	Slime bowl costume

Chapter 3.3 Debug flowcharts

Task A. Debug the flowcharts

Flowchart 1 - The error is the use of AND when it should be an OR

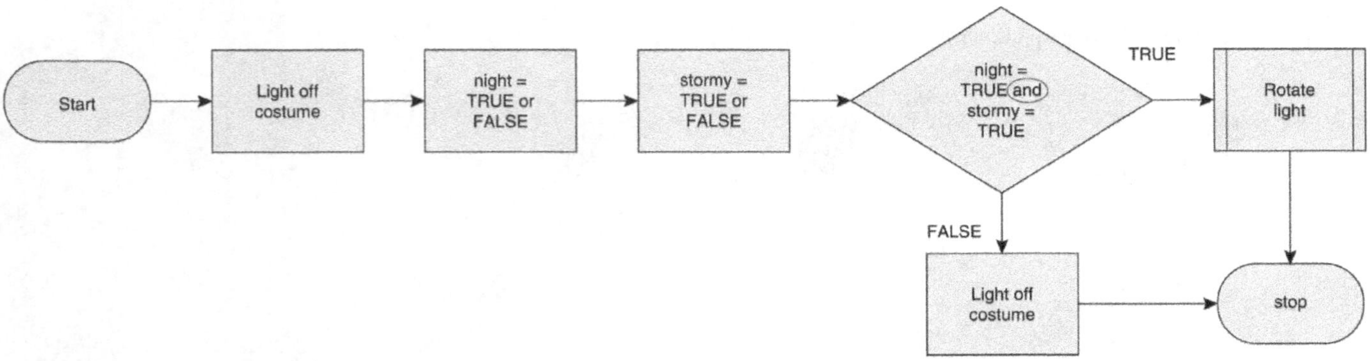

Correct flowchart

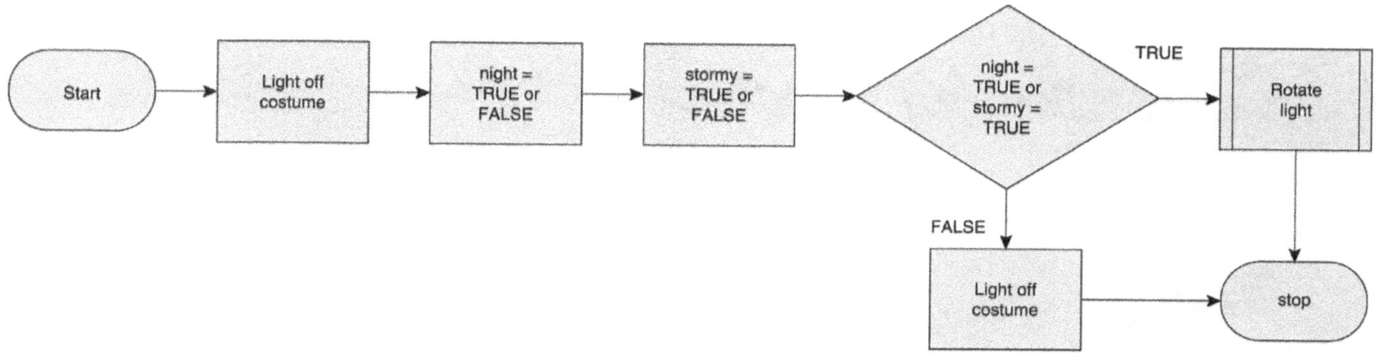

Flowchart 2 - The error is that both conditions need to be TRUE and it should use an AND. It is also missing which side is TRUE and FALSE.

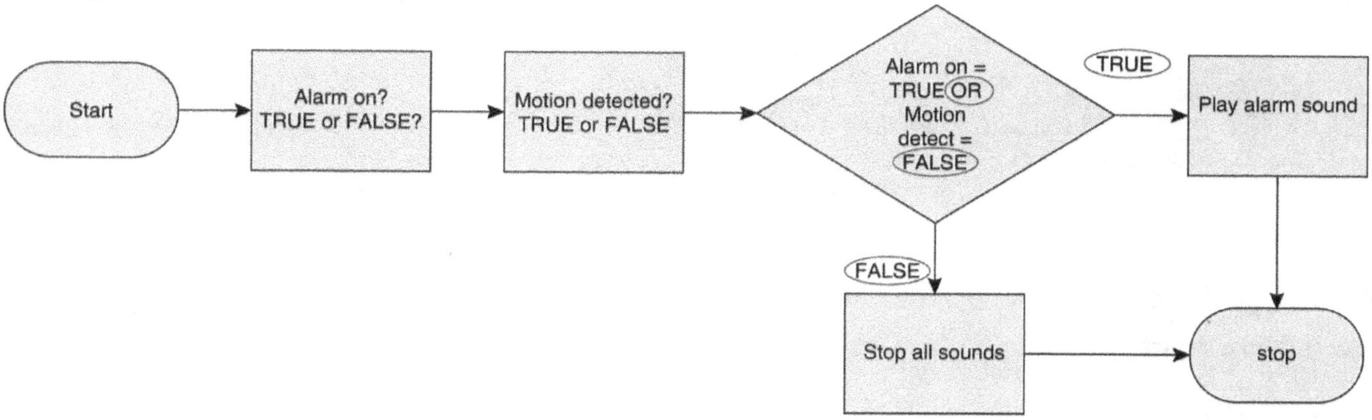

Correct flowchart

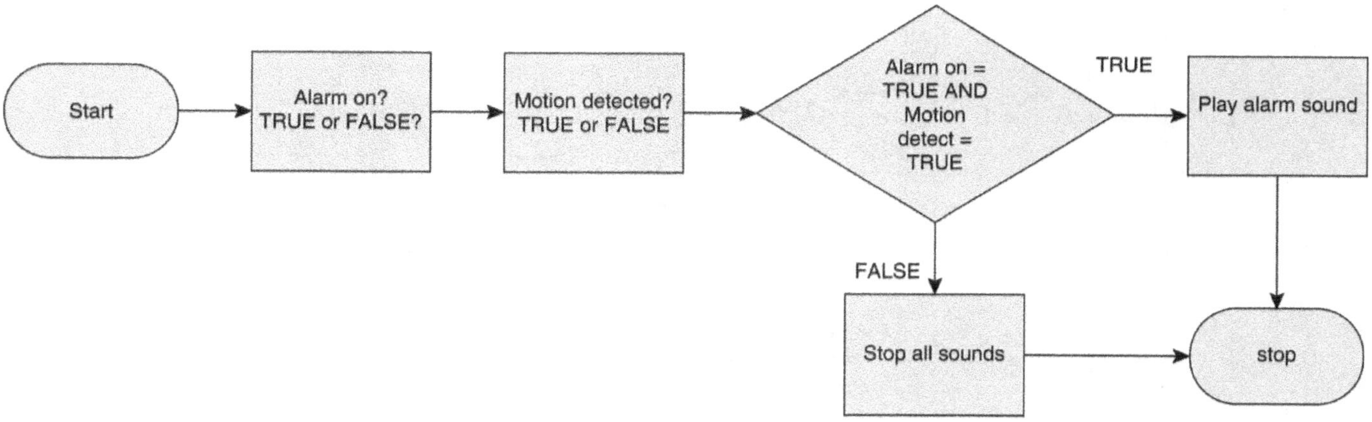

Flowchart 3 errors:

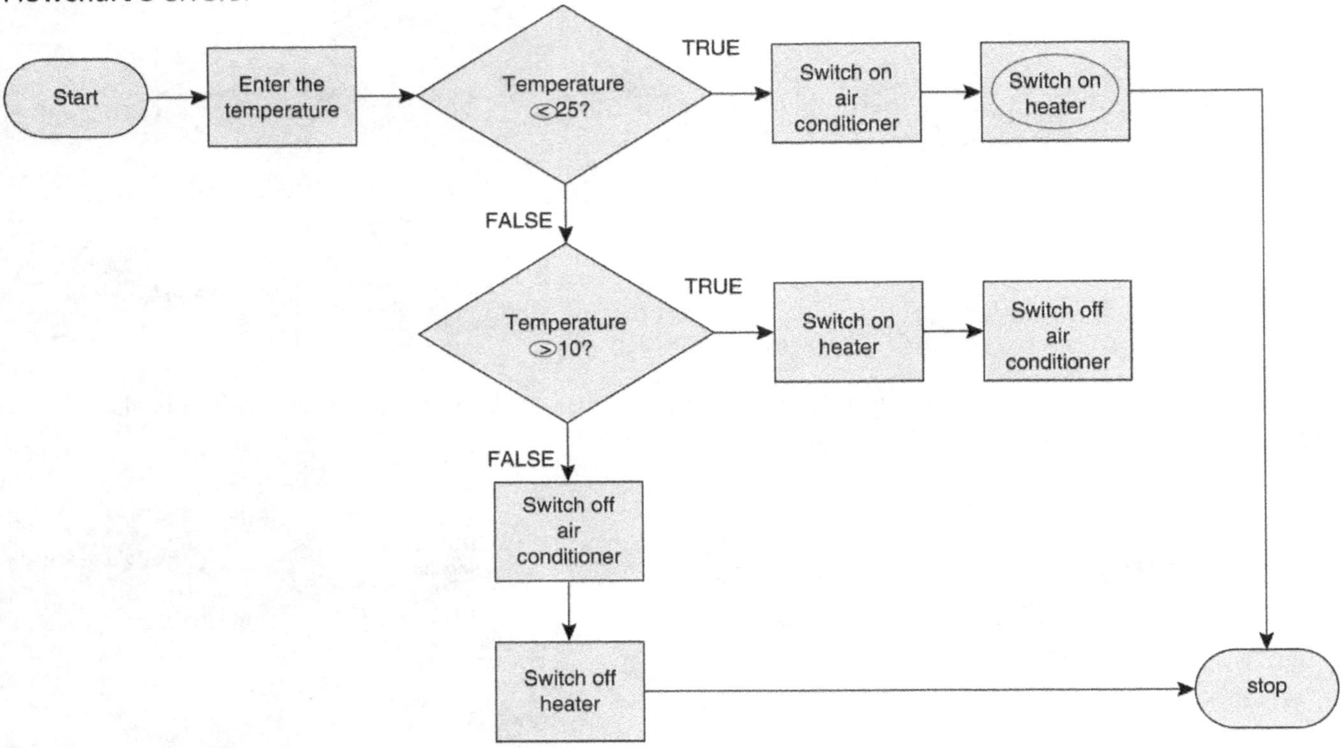

Correct flowchart

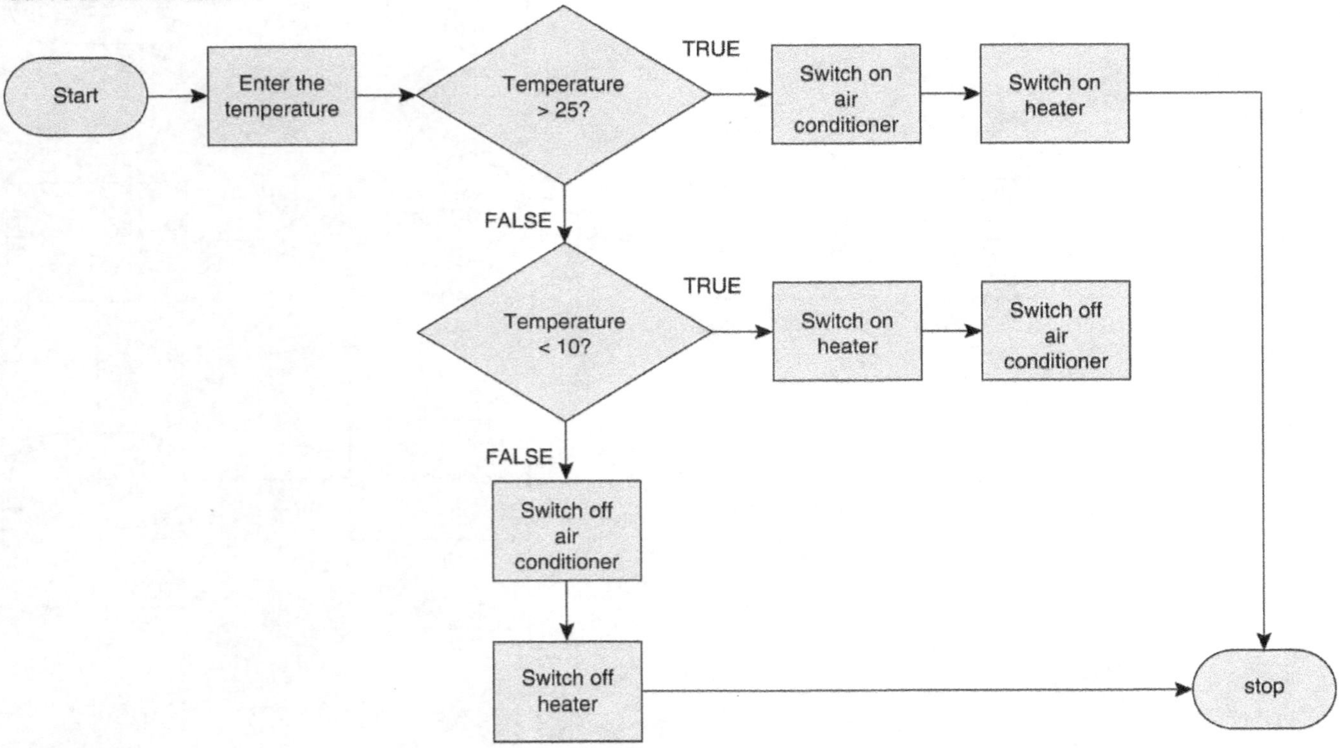

Homework

Two mistakes:

1. 'Tomato' should be 'Cucumber'
2. Instead of 'OR' it should be 'AND'

Chapter 4 – How computers work

Chapter 4.1 Software and automation

Task A. System or application software

Software name	System software?	Application software?
Word processor		✓
Antivirus	✓	
Block based programming software		✓
Online form creator		✓
Operating system	✓	
Photo editing software		✓
Web browser software		✓
A computer game		✓

Chapter 4.2 Binary representation

Task A. Count in binary

Decimal	8	4	2	1
0				0
1				1
2			1	0
3			1	1
4		1	0	0
5		1	0	1
6		1	1	0
7		1	1	1
8	1	0	0	0
9	1	0	0	1
10	1	0	1	0

Task C. Represent sound in binary

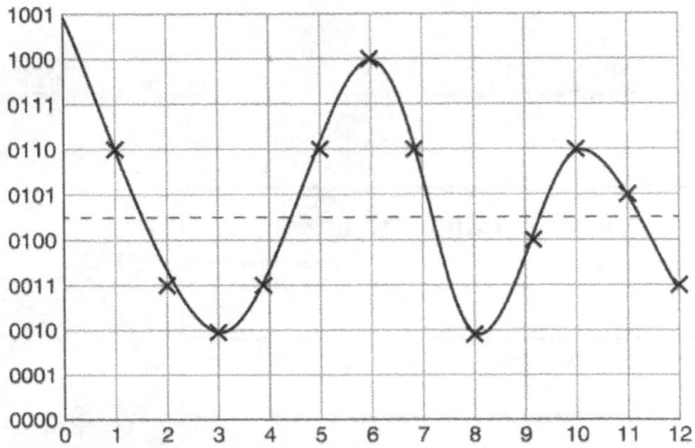

Time	Binary	Time	Binary	Time	Binary
1	0110	5	0110	9	0100
2	0011	6	1000	10	0110
3	0010	7	0101	11	0101
4	0011	8	0010	12	0011

Chapter 4.3 Represent images in binary

Task A. Draw the 1-bit colour depth images

(a)

1	0	1	0
0	1	0	1
1	0	1	0
0	1	0	1
1	0	1	0

(b)

1	0	1	0
1	0	1	0
1	0	1	0
1	0	1	0
1	0	1	0

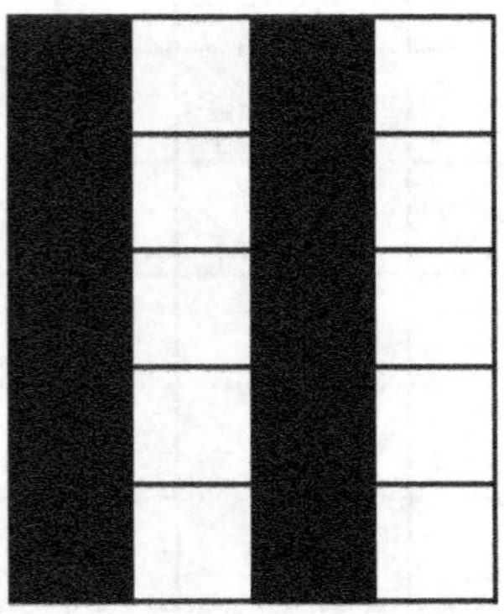

(c)

1	0	0	1
0	1	1	0
1	0	0	1
0	1	1	0
1	0	0	1

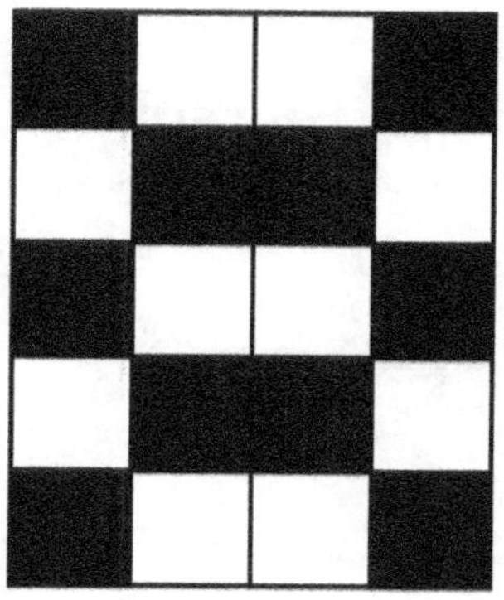

Task B. Colour combinations

1. The combinations of 1s and 0s that you can make with 3 bits are:
   010 011 100 101 110 111
2. With a 3-bit colour depth image, you can make 8 colours.
   The combinations of 1s and 0s that you can make with 4 bits are:
3. 0000 0001 0010 0011 0100 0101 0110 0111 1000 1001 1010 1011 1100 1101 1110 1111
   With a 4-bit colour depth image you can make 16 colours.

Homework

Spot the pattern

Number of bits	1	2	3	4
Combinations	2	4	8	16

The number of combinations doubles each time.
With 5 bits you can make 32 combinations.

Chapter 5 – Create with code 2

Chapter 5.1 Python: input and output

Task A. Print statements

- `print("Welcome to EduBlocks")`

Check for:
- Single or double quotes
- Round brackets
- print is in lowercase
- Text string matches exactly

Homework

Order the code

Code block	Number order
	2
	3
	1

Chapter 5.2 Python: Operators, conditions and data types

Task B. I need an int

1. Correct code:
   Answer: 1273941

2. Correct code:
   Answer: 40500

Task C. Using decimal numbers

An error will appear. 'invalid literal for int()'

Task D. Calculate ticket prices

Correct code:
   a. 18
   b. 33
   c. 114

Task E. True or False?

1. True
2. True
3. False
4. True

Homework

Match the Python blocks to their descriptions by drawing a line.

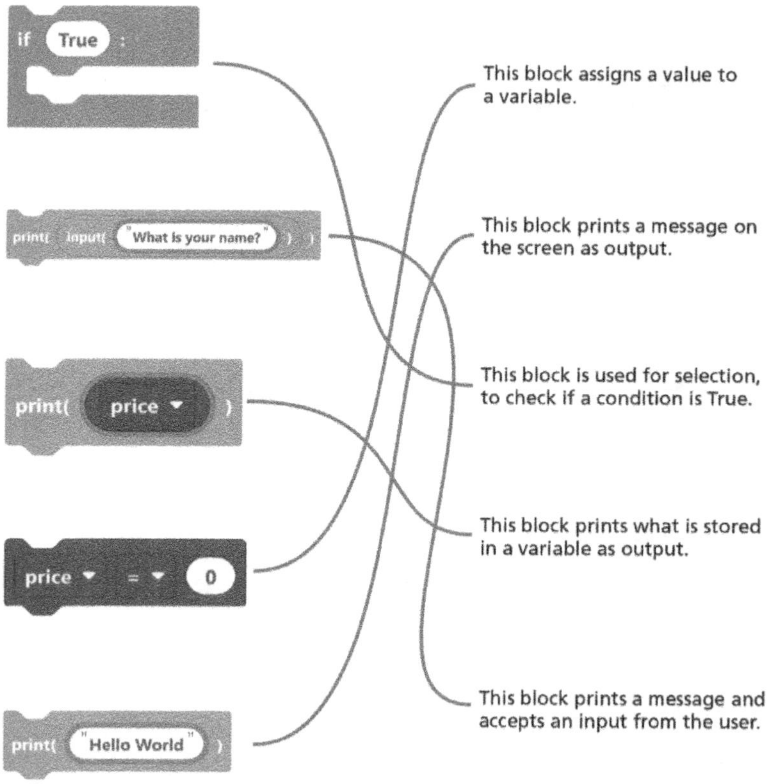

Chapter 5.3 Python: Boolean logic, comments and functions

Task A. Decisions

Recommendation	Age	Likes
The Girl Who Stole an Elephant	A number >= 8 such as 9	e
Maya and the Robot	A number >= 8 such as 10	r
Zola's Elephant	A number < 8 such as 7	e
Blippy the Robot	A number < 8 such as 5	r

Task B. Comments

Line numbers: 4, 7, 10, 13

Task C. Recognising patterns

The first and last print statements are exactly the same.

The second print statement is different. Some students may notice that part of the statement is the same; it's just the actual name of the book that is different.

Task E. Debugging a function call

The text 'I recommend' should be removed from the function call because it is always added inside the function.

Chapter 6 – Connect the world

Chapter 6.2 Wireless data transmission

Task A. Which wireless technology?

	Scenario	Answer
1	Wireless headphones	Probably Bluetooth (low cost, low power) but could be wireless with justification (streaming music from the internet)
2	Internet of Things device for monitoring moisture in the garden of a home	Typically Wi-Fi due to range and cost
3	Alert system for elderly person that sends a message	Cellular for reliability. Person may not have Wi-Fi or turn it off.
4	Robot programmable from an app	Probably Bluetooth but could be Wi-Fi, especially if the robot connects to the internet.
5	A washing machine that can be controlled and monitored from an app	Wi-Fi so you can monitor the washing machine when you are not at home.
7	A voice assistant device that can read audio books and control household devices using voice commands	Wi-Fi for connecting to the internet and to other smart devices. May also use Bluetooth to connect to a speaker or headphones. Extra credit if anyone know that other wireless protocols are also used for home automation.

Task B. 5G features

1	Jan wants to meet up with a friend during a busy celebration where lots of people are using their mobile phones in a small area.	Increased network capacity
2	Mobile phone usage has an impact on the environment due to power usage.	Improved energy efficiency
3	A sports fan wants their web page to load quickly so they can see the results.	Faster data transfer speeds
4	A medical device needs to provide accurate and timely data in an emergency situation.	Faster reliable communication for devices

Chapter 6.3 Encryption and data security

Task A. Decryption

1. **(a)** 4 **(b)** −2
2. THE PASSWORD IS LOVELACE

Task B. Encryption

1. **b**
2. **(a)** decryption **(b)** encryption

Task D. Explore the encryption algorithm

Answers could mention:

- The sprite moves to a start position in the top left corner.
- The code loops through the letters in the message.
- The number of repeats is the length of the message.
- The (next letter) variable corresponds to the current letter and is increased by 1 each time the loop runs.
- The costume name is the letter.
- The sprite switches to the costume corresponding to the current letter (using the costume name) and then moves to a new position.
- The sprite hides so you don't get an extra letter at the end.

Chapter 7 – The power of data

Chapter 7.1 Modelling data to make decisions

Task B. Data-driven decisions

1. **(a)** 0.30
   (b) 1.05
   (c) 81.24 (accept 82)

2. **(a)** 0.35
   (b) 0.83
   (c) 107.21 (accept 108)

3. **(a)** Radio station 1
   (b) They might not play the song as frequently in the future.

4. **(a)** Tab 'Station 1' cell 'B1'
   (b) 57.4 (accept 58)

Chapter 7.2 Organising and formatting data

Task A. Identify primary keys

1. **(a)** Repeats (not unique)
   (b) Empty cells
   (c) Repeats and empty cells

2. **(a)** Product ID
   (b) Registration no.

Task B. Highlight scores

3. 9

Chapter 7.3 Search and capture data

Task A. Searching databases

1. **(a)** Parrot, Penguin, Hummingbird
   (b) Penguin, Albatross
   (c) Parrot, Albatross
   (d) Penguin
   (e) None

Chapter 8 – Create with code 3

Chapter 8.1 Project plans

Task A. Order the tasks

Note that this is a model order. Students' answers may be different, in particular, in relation to when to set the budget and when to meet with the stakeholders.

Task	Order
Consider the design options for the app.	4
Build the app.	5
Test the app with students.	7
Meet with all stakeholders to decide on the scope of the project.	3
Test the app internally.	6
Establish a budget for the project.	2
Launch the app.	8
Create a timeline for the project.	1

Task B. Case study: Mobile ticketing

The main reasons the mobile ticketing failed were:

- missing stakeholders
- scope creep.

Task C. Effective design

1. b

Task A. Follow a test plan

#	Description	Input	Expected output	Result	Fix (if needed)
1	A welcome message displays	Switch on / play	A smiley face icon	Heart icon	Code should be `display.show(Image.HAPPY)` not `display.show(Image.HEART)`
2	User generates the first number	A	A random number between 10 and 100 is shown	5	Code should be `a = random.randint(10, 100)` not `a = random.randint (1, 10)`
3	User generates the second number	B	A random number between 10 and 100 is shown	1	Code should be `b = random.randint(10, 100)` not `b = random.randint(1, 10)`
4	The result displays	Shake	The result of adding the two numbers together is shown	4	Code should be `display.scroll(a + b)` not `display.scroll(a - b)`

Task B. Write and follow a test plan

#	Description	Input	Expected output	Result	Fix (if needed)
1	A welcome message displays	Switch on / play	"Hi" is scrolled	Hi	
2	User generates the first number	A	A random number between 1 and 12 is shown	12	
3	User generates the second number	B	A random number between 1 and 12 is shown	1	
4	The result displays	Shake	The result of multiplying the two numbers together is shown	144	The line of code should be `display.scroll(a * b)` not • `display.scroll(a * a)`

Chapter 8.3 Prototypes

Task A. Does it meet the requirements?

1. No, the requirements have not been met.
2. You can choose a costume but not the actual animal for the pet.
   There is no on and off button.
   There is an additional feature which is to play a sound, this is not mentioned in the brief.
   You could improve the design by making sure that you can choose the type of animal and by adding on and off buttons. The 'play a sound' feature is a useful addition and this should stay.

Websites

Introduction

Reference	Website URL
IntR1	https://edublocks.org/123-basic-set.pdf
IntR2	Classroom Talk and Computational Thinking | International Journal of Computer Science Education in Schools (ijcses.org)
IntR3	GBIC i4: Relevance evaluation report (w/ appendices 1-3) (teachcomputing.org)
IntR4	Guide+to+culturally+relevant+and+responsive+computing+in+the+classroom.pdf (raspberrypi.org)
IntR5	Fostering computational thinking through unplugged activities: A systematic literature review and meta-analysis | International Journal of STEM Education | Full Text (springeropen.com)
IntR6	Understanding Conceptual Transfer for Students Learning New Programming Languages | Proceedings of the 2020 ACM Conference on International Computing Education Research
IntR7	Teachers' Experiences of using PRIMM to Teach Programming in School | Proceedings of the 50th ACM Technical Symposium on Computer Science Education
IntR8	EduBlocks
IntR9	Welcome to the EduBlocks docs - Docs
IntR10	PowerPoint Presentation (edublocks.org)
IntR11	Scratch - Imagine, Program, Share (mit.edu)
IntR12	https://scratch.mit.edu/faq#accounts
IntR13	https://machinelearningforkids.co.uk/
IntR14	https://machinelearningforkids.co.uk/#!/help
IntR15	https://www.google.com/docs/about
IntR16	https://support.google.com
IntR17	https://www.google.com/sheets/about
IntR18	https://support.google.com
IntR19	https://www.google.co.uk/forms/about
IntR20	https://support.google.com
IntR21	https://microbit.org/get-started/getting-started/introduction
IntR22	https://microbit.org/teach/professional-development
IntR23	https://microbit.org/get-started/features/overview

Chapter 1

Reference	Website URL
C1L2R1	Scratch - Scratch Community Guidelines (mit.edu)
C1L2R2	https://scratch.mit.edu/community_guidelines
C1L2R3	Rules for using Raspberry Pi Foundation websites - Raspberry Pi Foundation
C1L2R4	https://forum.tynker.com/faq
C1L3R1	https://scratch.mit.edu/projects/923711779/fullscreen/
C1L3R2	https://scratch.mit.edu/projects/923711779/fullscreen/
C1L4R1	https://machinelearningforkids.co.uk/
C1L4R2	https://www.zooniverse.org/projects/aeuk/elephant-id

C1L5R1	https://machinelearningforkids.co.uk/
C1L5R2	https://scratch.mit.edu/projects/923711779/
C1L5R3	https://scratch.mit.edu/projects/923711779/

Chapter 2

Reference	Website URL
C2L1R1	https://docs.google.com

Chapter 3

Reference	Website URL
C3L1R1	https://scratch.mit.edu/projects/921858880
C3L2R1	The correct ingredients on Scratch (mit.edu)
C3L4R1	Rating Kiosk on Scratch (mit.edu)

Chapter 4

Reference	Website URL
C4L2R1	Count in binary on Scratch (mit.edu)
C4L4R1	https://coolors.co/
C4L5R1	https://www.pixilart.com/draw

Chapter 5

Reference	Website URL
C5L1R1	https://edublocks.org/
C5L1R2	https://edublocks.org/123-basicset.Pdf
C5L1R3	https://app.edublocks.org/project/LP83qy9v64WtGOR2HJzibnuFbkR2/s1wGmKmiveAeMuCXwBZF
C5L2R1	https://edublocks.org/
C5L3R1	https://edublocks.org/
C5L3R2	https://app.edublocks.org/project/bA8JOUKeaOMkt3GJiZLFuP57USZ2/Q0pyvcLAYEfHGt9Fjnyi
C5L3R3	https://app.edublocks.org/project/bA8JOUKeaOMkt3GJiZLFuP57USZ2/BZOkslBcww73zzXZzpma
C5L3R4	https://app.edublocks.org/project/bA8JOUKeaOMkt3GJiZLFuP57USZ2/Q0pyvcLAYEfHGt9fJNYI
C5L3R5	https://app.edublocks.org/project/bA8JOUKeaOMkt3GJiZLFuP57USZ2/BZOkslBcww73zzXZzpma
C5L4R1	https://app.edublocks.org/project/LP83qy9v64WtGOR2HJzibnuFbkR2/YO5cHjNrNEJHOTZDqJTy
C5L4R2	https://app.edublocks.org/project/LP83qy9v64WtGOR2HJzibnuFbkR2/YAvroSP48p0KphgefUD7
C5L4R3	https://app.edublocks.org/project/LP83qy9v64WtGOR2HJzibnuFbkR2/e0VVs66PU8OpYcnUpcEY
C5L4R4	https://app.edublocks.org/project/LP83qy9v64WtGOR2HJzibnuFbkR2/PJQz3MRCQuWSFO1AHdb1
C5L4R5	https://app.edublocks.org/project/LP83qy9v64WtGOR2HJzibnuFbkR2/YAvroSP48p0KphgefUD7
C5L4R6	https://app.edublocks.org/project/LP83qy9v64WtGOR2HJzibnuFbkR2/e0VVs66PU8OpYcnUpcEY
C5L4R7	https://app.edublocks.org/project/LP83qy9v64WtGOR2HJzibnuFbkR2/PJQz3MRCQuWSFO1AHdb1
C5L4R8	https://app.edublocks.org/project/LP83qy9v64WtGOR2HJzibnuFbkR2/YO5cHjrNEJHOTZDqJTy
C5L1R1	https://edublocks.org/

Chapter 6

Reference	Website URL
C6L3R1	https://scratch.mit.edu/projects/927514179/
C6L3R2	https://scratch.mit.edu/projects/927514179/
C6L4R1	https://scratch.mit.edu/projects/927518736/editor/
C6L4R2	https://scratch.mit.edu/projects/927518736/editor/

Chapter 7

Reference	Website URL
C7L1R1	https://scratch.mit.edu/projects/925090729/editor

Chapter 8

Reference	Website URL
C8L1R1	https://microbit.org/getstarted/features/overview/
C8L2R1	https://app.edublocks.org/project/GWomfRj1eHOwh9Qfd9KOjKfnWb82/eHs5FsyIMlNGatcu613p
C8L2R2	https://app.edublocks.org/project/GWomfRj1eHOwh9Qfd9KOjKfnWb82/8pjK9feEjQbnwtlj0C5y
C8L4R1	https://app.edublocks.org/project/GWomfRj1eHOwh9Qfd9KOjKfnWb82/0VtulULT77JpPT9vBV4l
C8L4R2	https://app.edublocks.org/project/GWomfRj1eHOwh9Qfd9KOjKfnWb82/mvFliepvbOw5CTLCu3Lk
C8L4R3	https://app.edublocks.org/project/GWomfRj1eHOwh9Qfd9KOjKfnWb82/QiNrsANfl4SnAFecqkaR
C8L4R4	https://app.edublocks.org/project/GWomfRj1eHOwh9Qfd9KOjKfnWb82/0VtulULT77JpPT9vBV4l
C8L4R5	https://app.edublocks.org/project/GWomfRj1eHOwh9Qfd9KOjKfnWb82/mvFliepvbOw5CTLCu3Lk
C8L4R6	https://app.edublocks.org/project/GWomfRj1eHOwh9Qfd9KOjKfnWb82/QiNrsANfl4SnAFecqkaR